STRATEGIC MINERALS

STRATEGIC MINERALS
The Geopolitical Problems for the United States

Ewan W. Anderson

PRAEGER

New York
Westport, Connecticut
London

Library of Congress Cataloging-in-Publication Data

Anderson, Ewan.
Strategic minerals: the geopolitical problems for the United States / Ewan W. Anderson.
p. cm.
Bibliography: p.
Includes index.
ISBN 0–275–93062–9 (alk. paper)
1. Mineral industries—Political aspects—United States.
2. Strategic materials—Government policy—United States. 3. United States—Foreign economic relations. I. Title.

HD9506.U62A726 1988
333.8'5'0973—dc19 88–12028

Library of Congress Catalog Card Number: 88–12028
ISBN: 0–275–93062–9

First published in 1988

Praeger Publishers, One Madison Avenue, New York, NY 10010
A division of Greenwood Press, Inc.

Printed in the United States of America

The paper used in this book complies with the Permanent Paper Standard issued by the National Information Standards Organization (Z39.48–1984).

10 9 8 7 6 5 4 3 2 1

To

MLA and WWA

Contents

Figures and Tables

FIGURES

TABLES

Preface

Strategic minerals are by definition vital, and potential problems of supply maintenance have exercised the policymakers of many governments. However, policymaking in this sphere presents problems of a complex nature. The subject is essentially multidisciplinary, including aspects of geology, technology, economics, and politics with geopolitics. To obtain an accurate perception, vital for policy formulation and implementation, thus presents great difficulties.

The aim of this study is to assess all key aspects of the strategic minerals vulnerability of the United States. This involves an examination not only of the main strategic minerals, but also of the risks attendant on obtaining them from certain sources and on transporting them by certain supply routes. This is the first attempt, in one volume, to analyze in detail the total global strategic minerals infrastructure of the United States. It is considered therefore that the book should be of interest to those studying a wide range of subjects from politics to geography.

Once the various vulnerabilities have been highlighted and assessed, potential government responses can be considered. However, which responses are implemented as policy depends very much on the data reaching the decisionmakers. This book and its conclusions are only one data source, albeit, it is hoped, an objective one. Other sources available to the United States government and the processes of policy formulation and implementation are considered in a companion volume: *The Structure and Dynamics of United States Government Policy-making: The Case of Strategic Minerals* (Anderson, 1988).

Data for both these studies were collected over a nine-month period in the United States, during which I worked as a consultant to a number of institutions, including two government departments, an independent research organization, and a trade association. For a detailed examination of vulnerability, a risk assessment model was designed and tested specifically for this study. Information and ideas were checked in a series of interviews with key officials in all the major departments and agencies of the government concerned with the strategic minerals issue.

While the disparate range of variables involved in the strategic minerals problem render the issue challenging in its complexity, it is for this very reason that the subject itself is never lacking in interest. It is hoped therefore that this book not only will be useful to policymakers, but also will include topics of interest to all readers, whatever their background or persuasion.

The research for this book was made possible by a leave of absence, granted by the University of Durham. In the United States, support was most generously provided by the National Strategy Information Center and by the American Mining Congress, neither of which body attempted, in any way, to influence the outcome of the work. Of crucial importance also was access to the thinking of the National Materials and Minerals Program Advisory Committee (Department of the Interior) to which I was appointed consultant. These various bodies, together with all the departments and agencies of the United States government, which so freely and cheerfully provided information and advice, are gratefully acknowledged.

On a personal level, the contributions of Robert Silano and my house colleagues in Arlington were invaluable. I am also particularly indebted to Greg Anderson for his good-natured and skillful help with editing the book, Ottie O'Brien for her careful indexing, and Bridget Roberts, Catherine Reed, and Margaret Bell for all their hard work in typing the manuscript.

STRATEGIC MINERALS

1

The Strategic Minerals Problem: Introduction

GEOPOLITICS AND THE GLOBAL POSITION OF THE UNITED STATES

During the course of the twentieth century, geopolitics has varied considerably in popularity, respectability, and even definition. The term was probably first used by Kjellen[1] (1899) to describe "the science of the state as a realm in space." The major merit of such a definition is that it is neither deterministic nor prescriptive. However, in the modern setting, Kjellen's definition is limited because it carries no suggestion as to its application. In the light of modern developments, S. Cohen's definition (1973, 29) is more appropriate: "The relation of international political power to the geographical setting." Since it provides geographical material for political decisions, geopolitics is essentially a practical subject, and therefore the use of the term "political power" is particularly relevant. However, it can be argued that the epithet "international" is unnecessarily restrictive. A fundamental premise of geography is the interdependence of the total global environment. Thus factors at the very local level may influence international decisions; conversely, global politics may exercise a strong influence at the local level.

This idea obtains quite clearly in the study of strategic minerals, since variables affecting a mining site may acquire significance of global proportions. Therefore, while the accent must be on the international or global level, it must be understood that interaction can pervade the entire environmental system. It is probably because geopolitics introduced (or more correctly reintroduced) geographical thinking on a global scale that this accent has persisted. There is no logical reason that geographical influences upon local political decisions should not be considered as an integral part of geopolitics. Thus it is contended that geopolitics is not scale-specific, and the interaction of inputs, political and geographical, must be considered on all scales.

Nonetheless, the development of the subject has been dominated by thinking on the global level, the scale on which key modern proponents such as A. T. Mahan (1890), H. J. Mackinder (1904, 1919, and 1943), N. J. Spykman (1944), S. Cohen (1964 and 1973), and R. E. Walters (1974) have worked. Furthermore, the major thrust in nearly all of these studies has been military, and thus they represent only one facet of the subject. In resource geopolitics, military action represents the ultimate escalation of disagreements and hostilities, political, economic, cultural, and even historical.[2]

The present danger to the United States resulting from strategic material dependence must be envisaged in its geopolitical setting. This setting consists of actors of different types and sizes, interacting within a particular environment, which also varies in type and scale, to produce a constantly changing series of effects. These effects feed back to and, in their turn, influence both the actors and the environment. Thus it is an extremely complicated network, and arbitrary boundaries must necessarily be drawn to allow analysis.

Changes in the geopolitical setting impinge upon U.S. national security, which by its very nature is resistant to simple definition (e.g., Trager and Kronenberg 1973; Trout and Harf 1982). In the present context, the working definition given by F. N. Trager and F. L. Simonie (1973, 36) seems appropriate: "National security is that part of government policy having as its objective the creation of national and international political conditions favourable as to protection or extension of vital national values against existing and potential adversaries." With the key role of the government being thus the protection of national values and interests, an adequate defense base is a fundamental requirement.

On September 11, 1935, Sir Samuel Hoare, the British foreign secretary, in the course of a famous speech at Geneva, stated as follows:

> Abundant supplies of raw materials appear to give peculiar advantage to the countries possessing them. It is easy to exaggerate the decisive character of such an advantage, for there are countries which, having little or no natural abundance, have yet made themselves prosperous and powerful by industry and trade. Yet the fact remains that some countries, either in their native soil or in their colonial territories, do possess what appear to be preponderant advantages; and that others, less favoured, view the situation with anxiety (Plummer 1937, 9).

It should be even more obvious now than it was 50 years ago, when these words were spoken, that national security is dependent not only upon a country's military might, but also upon the strength of its economy. It is generally accepted that, in the event of war, the nation with a well-developed defense and industrial base and the ability to mobilize (e.g., Merritt and Carter 1985) its full economic potential is likely to prevail over its adversaries. In many ways, however, lines between war and peace are now so blurred that the conditions of armed conflict, international crisis, and intense economic, military, and ideological competition tend to blend into one another. Thus national security has come to be seen as having distinct military, politico-ideological, and economic connotations (e.g., Seabury 1983).

Cast in somewhat different terms, a symbolic relationship may be said to exist between economic prosperity and national security. A strong and thriving economy fosters research and development, which in turn gives rise to sophisticated military requirements and makes a relatively high level of defense spending possible. Furthermore, during a period of rising tension or war, the industrial base of such an economy would be expected to transform itself rapidly into a war economy.

At the same time, a country that possesses a well-structured, multifaceted defense establishment can also provide security for its allies and trading partners while protecting vital lines of communication. If the United States were to falter in its international commitments, essential overseas connections would be jeopardized and perhaps even lost. Simultaneously, the economy would be denied certain strategic resources and decline, with the net result that defense production capabilities would suffer. National security would then face a vicious, descending spiral that would gravely undermine the readiness of the armed forces and the domestic economy.

Among the actors on the world stage of strategic material importance are the superpowers, the United States and the Soviet Union. While the United States is markedly import dependent, the Soviet Union is notably self-sufficient. The People's Republic of China, by all accounts very rich in minerals, may well disturb the present bipolar position in time. Among the mesopowers, South Africa, Australia, and Canada in particular, but also countries such as Brazil and India, are important. Increasingly, however, micropowers such as Gabon, Guinea, and even perhaps New Caledonia are becoming significant on a global level. Nonterritorial bodies such as the mining and materials multinational corporations and multilateral intergovernmental bodies such as the United Nations Conference on Trade and Development (UNCTAD) are also important. The changing pattern of relationships among all of these within a pervasively competitive global environment (sometime of a hostile nature) has a direct impact on access to minerals and affects the status of both supplier and consumer.

Thus, at a time when competition for resources is keen, should a producer for some reason decide to reorient its trade away from the United States, a number of consequential effects would follow. The influence of the United States is likely to be undermined in the country concerned and perhaps in the region. The trade pattern affecting the U.S. economy would register changes of a potentially serious nature, with increased vulnerability a probable consequence. It is therefore important to examine the complete infrastructure at every point from the actual extraction onwards, through the transport network and the exit points of the source country to the lines of communication leading to the United States.

MINERAL DEPENDENCE AND VULNERABILITY

In many of the arguments advanced about nonfuel minerals, import dependence is taken to be synonymous with vulnerability. If a country has to import a high proportion of a particular mineral, then clearly its supply is fundamentally vulnerable to disruption. However, this may range from fairly minor to potentially catastrophic disruptions depending on a variety of other factors. Alarms about dependence have arisen at various times since World War II, but undoubtedly the major initiative leading to the present debate can be located in events that took place in the Organization of Petroleum Exporting Countries (OPEC) during the years 1973–74. Sudden increases in the price of petroleum, set by OPEC, affected almost all parts of the world. Import dependence suddenly became headline news. More economic forecasts (e.g., those produced by the Club of Rome, see Meadows et al. 1972) pointed to problems of domestic resource depletion. Added to this, anxiety became widespread concerning the attitude of many of the developing supplier countries.

In an effort to achieve a degree of congruity between political and economic independence, Western assets were nationalized or expropriated or seized, sometimes summarily, in several countries. Where this led to a marked deterioration in mining efficiency, import vulnerability obviously increased. At the same time, the developing countries became increasingly articulate within international bodies, demanding a fairer share of world resources under a "New International Economic Order" (NIEO). One result of all this was that the terms "dependence" and "vulnerability" came to be used more or less interchangeably.

From a somewhat different perspective it can be appreciated that world trade is based on mutual interdependence and even the best endowed country can only be fully self-sufficient at some cost. Such a cost is an increasing rate of inflation, accompanied by a decreasing rate of economic growth. An expanding economy requires a rising volume of raw materials, and if these can be imported more cheaply than they can be produced at home, the result could be described as beneficial. Furthermore, as many of these suppliers are also developing countries, they can obtain foreign exchange needed for imports from the developed world. This line of reasoning generally follows the trend set by the Brandt Report (Brandt et al. 1980).

Although such a line of reasoning is broadly accepted, there is a school of thought according to which certain materials are so vital to the well-being of the United States that their supply must always be guaranteed in order that any threat from outside could be met without delay. Strategic reliability includes a number of considerations, in particular the critical nature of the material, the number and reliability of alternative sources, and possibilities for recycling, substitution, and conservation (e.g., Advisory Group for Aerospace Research and Development [AGARD] 1984). Furthermore, the entire supply infrastructure needs to be kept under close observation with a specific view to preventing its disruption.

Thus dependence can, in certain cases and under certain conditions, be equated with vulnerability. It is difficult to achieve a balance between ensuring conditions under which free trade can flourish and making sure that defense is not placed at risk. As the report entitled *Strategic and Critical Nonfuel Minerals: Problems and Policy Alternatives* (Congressional Budget Office 1983, 3) concludes:

> Relying on imported supplies of materials poses both costs and benefits. The costs of such reliance consist of risks that shortages may occur as a result of military, political, or economic contingencies, or natural disasters in the country of origin. The benefits consist of lower costs for defense and industrial production which, in turn, lower costs to United States consumers and make United States products more competitive in domestic and foreign markets.

Policies aimed at alleviating the problems arising from dependence on foreign mineral suppliers must take account of these costs and benefits.

STRATEGIC MINERAL RESOURCES

The importance of access to resources is recognized at the highest levels within the United States government. It has been said that no other issue has been debated so often and discussed at such length without an agreement being reached. Five major studies deal with the problem, each published at a crisis point.[3] Produced under different circumstances, these reports identify a number of issues in common. The recommendations have invariably tended to be in favor of ensuring free trade, underlying which is a strong belief that private enterprise will triumph in the end. For example, the report of the commission of 1952 (known as the Paley Commission) concluded that (p. 21): "the United States must reject self-sufficiency as a policy and instead adopt the policy of the lowest cost acquisition of materials wherever secure supplies may be found: self-sufficiency, when closely viewed, amounts to a self-imposed blockade and nothing more."

The same policies were reaffirmed in 1976 when the National Commission on Supplies and Shortages (NCSS) concluded (p. 31): "Import dependence ... is not a problem as such; for example, it is simply misleading to cite a United States 'deficit' in the minerals trade as though this represented in some pertinent sense a national loss."

Recent policy statements have adopted a somewhat different stance. This is evidenced, for example, in the *National Materials and Minerals Program Plan and Report to Congress* (April 1982) issued by the White House. In his letter of transmittal, the president recognized four key aspects of a sound national minerals policy:

(a) the critical role of minerals to our economy, national defense and standard of living;

(b) the vast unknown and untapped mineral wealth of America and the need to keep the public's land open to appropriate mineral exploration and development;

(c) the critical role of government in alerting the Nation to minerals issues and in ensuring that national decision-makers take into account the impact of their decisions on minerals policy; and

(d) the need for long-term, high potential payoff research activity of wide generic application and to improve the augmented domestically available materials.

This policy initiative met with a mixed reception. Implementation has been slow, although the mining industry has given general approval to the new policy on the basis of the following key points.[4]

(a) the withdrawal of public lands from minerals exploration and development;

(b) the increasingly dangerous dependency of the United States on vulnerable foreign sources of minerals of strategic importance;

(c) the lag in establishing a legal regime to foster deep-seabed mining; and

(d) the precarious state of the nation's strategic material stockpile.

Clearly evident in the policy statement were the two main strands of the argument. On the one hand, it is necessary to maintain a free market initiative; on the other, the vulnerability of the United States to the supply of minerals from insecure sources must be somehow reduced. Furthermore, the establishment of the Cabinet Council for Natural Resources and the Environment (CCNRE) meant that the coordination and implementation of minerals and materials policy would be closer to the president than at any time since the end of World War II. At the same time, the policy statement and the subsequent debate led to an increase in the public awareness of the importance of strategic minerals for the United States. It also reinforced the publicly expressed anxiety about the undeclared "resource war" that originally surfaced during the late 1970s when the prices of many key nonfuel minerals peaked (Figure 1.1). The two events, the resource war on the one hand and the rise in prices on the other, of course were not unrelated.

Although it can be argued that market prices quoted refer only to specific forms of each mineral, thus rendering their function as a guide to the changing strategic status of particular materials problematic, the price of the metal used in industry and its price in commodity exchange are still related. Furthermore, prices are readily obtainable and encapsulate a range of factors, geological, political, and economic. This period of price peaking (1978–81) occurred at a time when political instability was rife in several parts of the world. The insurrection in Shaba (Zaire) and the intervention of external forces accompanying it gave rise to a sudden and massive

Figure 1.1
Price Fluctuations

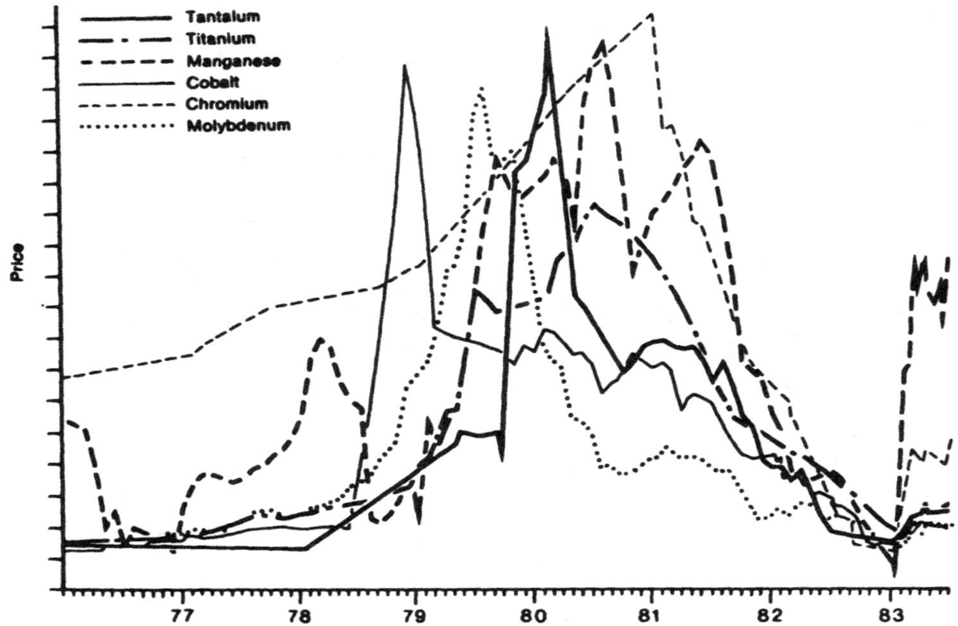

Source: Advisory Group for Aerospace Research and Development, North Atlantic Treaty Organization (AGARD/NATO), CP 356—Materials Substitution and Recycling in April 1984.

increase in the price of cobalt, which in turn affected the steel industry. A certain amount of panic buying was set off by these developments. The fall of the shah of Iran and the second round of oil price increases focused attention on raw materials for the second time since major intervention in 1973. During this period the Grand Mosque in Mecca was temporarily occupied, President Sadat was assassinated, and the first Reagan administration was elected. The superimposition of a large-scale rearmament program in the United States upon this somewhat unstable world background had a sharp impact on the prices of strategic minerals.

Since then, in the wake of the worldwide recession that has hit industry in general and the steel industry in particular, the cost of minerals has fallen and there has been a glut in oil. With the development of non–Middle Eastern sources of petroleum (for example, Alaska and Mexico), it seems unlikely that the collective strength of OPEC, at least in the medium term, will regain the devastating influence it exerted during the 1970s.

KEY TERMINOLOGY IN THE DISCUSSION

In discussing the vulnerability of the United States and the supply of minerals, the epithets "critical" and "strategic" are commonly used. The first is generally taken to relate to the importance of the end use of the minerals and the second to the risks involved in securing their supply. However, there is no clear-cut definition of what constitutes a strategic or critical mineral. The Strategic and Critical Materials Stock Pile Revision Act (1979) has combined the two definitions in referring to materials that (a) would be needed to supply the military, industrial, and civilian requirements of the United States during an emergency;[5] and (b) are not found or produced in the United States in sufficient quantities to meet such need.[6]

The criteria for the inclusion of minerals in this category embrace possibilities for substitution, conservation and re-cycling, the prospects of increasing domestic production at reasonable prices, the percentage of import dependence upon insecure sources, and the scale of economic damage that would result from a disruption of their supply.

By contrast, the Office of Minerals Policy and Research Analysis (Department of the Interior) (OMPRA), in developing a critical minerals index (Adams, White, and Grichar 1979), has drawn attention to two factors in particular that determine the criticality of a mineral (pp. 1–2):

(a) the likelihood that the future supply of the particular mineral to the U.S. might be interrupted.

(b) the importance of the mineral to the functioning of the U.S. economy.

OMPRA discounted the commonly accepted view that an indication of the potential mineral supply problems would be contained in the estimated percentage of consumption of a specific mineral by the United States through import as both simplistic and misleading.

The Office of Technology Assessment (OTA) report entitled *Strategic Materials: Technologies to Reduce U.S. Import Vulnerability* (1985) has employed the term "strategic" to cover both criticality and vulnerability. According to the report, criticality is measured by the proportional requirement in applications essential to the United States, both civilian and military; vulnerability is assessed according to the risk of supply interruption, together with the scale and duration of the potential interruption. A strategic material may therefore be defined as "one for which the quantity required for essential civilian and military uses exceeds the reasonably secure domestic and foreign supplies and for which acceptable substitutes are not available within a moderate period of time" (p. 11). The report therefore followed the trend of recent literature that has combined all aspects of criticality, vulnerability, and dependence into the single term "strategic." For the purposes of the present study, in order that a mineral might be designated as "strategic," the following three criteria, in the order of importance listed, must obtain (Anderson 1984, 2-1):

(a) the resource is critical for defense and defense-related industries;

(b) there is a marked degree of import dependence; and

(c) there are few significant sources of supply.

If the first two of these criteria are met, their supply might still present problems. If all three criteria are fulfilled, then the likelihood of potential geopolitical difficulties would have to be taken seriously if a resource crisis is to be avoided.

The position is not static and could be radically altered by a number of events. For example, new discoveries of mineral reserves might increase the number of significant sources of supply. With the development of deep offshore mining, probably sometime during the early part of the next century, the pattern of mineral production is likely to undergo major changes. Political decisions or changes in economic circumstances may render mining in the United States, now considered to be uncompetitive for many key minerals, commercially viable. Developments in materials engineering and other technological innovations could obviously affect the cost of domestic mineral recovery, thereby giving rise to differences in the pattern of criticality between the defense industry and other sectors of the economy.

While it is possible to speculate on a number of other trends and their likely effects, it must be borne in mind that the evolution of all change, political, social, economic, technological, and scientific, takes time. For example, the lead time for initiating ocean mining is expected to be in excess of 20 years, whereas a new mine and its infrastructure could probably be developed in ten to twenty years. The incorporation of substitutes in sensitive technologies can be achieved in approximately five years, whereas a processing plant can be reopened in about two years. Therefore, future planning is not the only solution to problems of mineral availability; it is also necessary to safeguard existing production capabilities and to guarantee access to foreign sources of minerals for some time to come.

Changes facilitating a decrease in dependence on foreign sources of nonfuel minerals are likely to be restricted by such factors as limited funding for research and development, lack of market incentives, and industrial inertia. In order to overcome these obstacles, the U.S. business community must become fully aware of the situation so as to be prepared for national emergencies that might lead to a resource crisis. In the short term, this might necessitate measures that would increase costs, but the result in the long term would be a greatly improved position for the industrial sector and the economy as a whole. For example, a technical stockpile of research and development data could provide an invaluable complement to the national stockpile of strategic minerals.

In order to appreciate the implications of the problems of resource availability to an adequate extent, the four following questions can be posed.

Is the Mineral Critical?

Because the level of criticality depends on the particular properties of a mineral, what is the potential for substitution? Could vulnerability be reduced by improved conservation or recycling techniques? A great deal of expertise has already been developed in these fields, and an ideal opportunity exists now when supplies are not threatened for a reappraisal of the means available for reducing vulnerability.

Is the United States Import Dependent?

Can domestic supplies of primary or recycled minerals be used to meet industrial resource requirements? Among other resources, the United States does possess domestic sources of four of the most obviously strategic nonfuel minerals: manganese, chromium, cobalt, and platinum-group metals. Current market prices do not make their production commercially viable. Yet, with price adjustments and perhaps a guaranteed floor price, there is no reason the United States should not become, as it once was during the 1960s, a producer of these resources. In order for domestic mining in the United States to become viable, the world price of minerals should be multiplied by factors as shown below:

Manganese	× 5.0
Chromium	× 5.0
Cobalt	× 2.5
Platinum group	× 1.2

With regard to the platinum group, this has become the case in actual fact since 1986. Business and industry should cooperate more effectively with the federal government in this area to ensure a stable supply of minerals in the event of imports being disrupted.

Is There a Concentration on Only a Few Sources of Supply?

Is it possible to diversify those sources? When there are only a few suppliers, there may be no escape from concentration, but, on the other hand, reliance on foreign supplies may occur for other reasons such as historical connections or the quality of the product. If the number of suppliers is restricted, short-term dependence may be a problem. If the number of current world products is limited, there might be at least medium-term problems. Longer-term reliance would depend on the disposition of known world reserves. Furthermore, given the complexity of world trade patterns, secondary dependence may occur when imports are acquired from one country that, in turn, obtains them from another. For example, while cobalt may be purchased directly from Belgium/Luxembourg, the true reliance is upon Zaire, the original supplier of the mineral. Thus if the industrial sector is a large-scale user of a particular mineral or world production is severely restricted, the problem of dependence is likely to be acute.

Are There Geopolitical Problems?

If there is import dependence, geopolitical problems are likely to arise; and, if there is a concentration of imports from only a few sources, there is an even greater risk that geopolitical problems could disrupt mineral supplies. Essentially, the geopolitics of resources results when consuming nations are not also the producers. Therefore a wide range of factors, of which several are subjective in character, is involved, varying in importance from the global level to the local level.

The designation of a mineral as strategic thus depends on geological, technological, economic, and geopolitical factors. Broadly speaking, the first three can be defined in reasonably objective terms. Therefore, in any evaluation, the fourth, the geopolitical factor, becomes crucial. In order to minimize subjectivity, it is necessary to enter the field of risk analysis.

THE UNITED STATES STRATEGIC MINERALS POSITION

Whereas the threat in 1980 was posed in the form of unrecognized but nevertheless serious problems arising out of a disruption of nonfuel mineral supplies, today a real danger may be the complacency that has settled over the issue of mineral dependence. Even those who heeded the warning tend to dismiss the earlier stance as alarmist or, in any event, to assume that the likelihood of a resource crisis is more remote now than a few years ago.

In the following discussion on the strategic mineral vulnerability of the United States, it will be demonstrated that this threat is still present and has, in some ways, intensified during recent years. This argument is reinforced by a consideration of international trading patterns and a global infrastructure that shows that the U.S. economy has become more, rather than less, dependent upon foreign sources of minerals.

Indeed, the strategic minerals situation may even have been exacerbated by the aid policy of the United States government, which sustains uneconomic mining in a number of areas of the world to the detriment of the long-term development of the U.S. domestic mining industry. Furthermore, political changes in many of the developing supplier countries had led to a loss of efficiency in production and increased the threat to overall stability. However, this may vary from region to region within a country and even from mine to mine; detailed risk assessments would therefore be necessary for gauging the size and intensity of the problems thus caused.

Such an adverse trend is further accentuated by the decline of the United States merchant marine, which in many respects is now a minor component of the world fleet. Thus the decline of the U.S. mining and processing capability, the reduction of the U.S. maritime freight capacity, and the intensification of conditions of political instability in certain countries of the Third World, taken together, must give cause for concern. Unless this trend is reversed, the situation can only get worse during the coming decades. Because the interlocking elements in the infrastructure are complex, appropriate solutions can only be produced, in the domestic context, by the government in close cooperation with the business sector.

The present lull in demand for strategic minerals provides a temporary but important opportunity for addressing some of the perennial problems of resource availability and for evolving appropriate remedial procedures and actions. The private sector, in collaboration with federal preparedness planners, must bcome more aware of those difficulties, actual and potential, that affect the national mineral producing and refining capabilities. Government and industry would, therefore, have to develop jointly emergency measures designed to help reduce the vulnerabilities inherent in the current pattern of dependence on foreign supplies of resources. Such vulnerabilities could lead to situations that might pose a serious threat to the capacity of the U.S. industrial base, not only to the meeting of military requirements, but also to the production of essential consumer goods.

These issues affect all individuals because they have an important bearing on everyday life. In an emergency or period of tension, military requirements would come first; but the maintenance of the economic and social fabric of society must be of almost equal importance. The strength of the country cannot be measured only in terms of the number of trained military personnel and the amount of sophisticated hardware, but is crucially influenced by the will of the people, which in turn would depend on how and at what level the society is maintained. This is both a quantitative and a qualitative question.

Issues such as those relating to the declining industrial base of the United States, the drastic reduction of the merchant marine, and the correct use and composition of the stockpile can be influenced by the business community, but they are also susceptible to public pressure. Thus the ordinary person in the street is not crucially affected by decisions reached as a result of the present discussion, but can and indeed should enter into the arguments. In order to do so, it is vital to be informed.

NOTES

1. According to Muir (1975, 5) who refers to Kjellen (1917).

2. This is succinctly discussed in, for example, Gray (1977) and Zoppo and Zorgbibe (1985).

3. See *The President's Materials Policy Commission Report* (1952); *The Public Land Law Review Commission Report* (1970); *The National Commission on Materials Policy Report* (1973); *The National Commission on Supplies and Shortages Report* (1976); and *The Non-Fuel Minerals Policy Review Report* (1979).

4. American Mining Congress (AMC) 1982, internal paper.

5. The term "emergency" means a general declaration of emergency with respect to national defense made by the president or Congress.

6. Federal Emergency Management Agency (FEMA) 1986.

2

Strategic Minerals: Dependence

In 1984 the U.S. economy, with a gross national product (GNP) of $366 trillion, required processed nonfuel minerals to a value of $253 billion. These resources were derived from three sources: $23 billion in domestically produced minerals and $5 billion in imported unprocessed minerals. In addition to these, the United States imported $34 billion worth of processed nonfuel minerals. The combined value therefore of imported nonfuel minerals, both processed and unprocessed, amounted to $39 billion, which was over 50 percent of the total value of imported crude oil and $19 billion more than the value of U.S. mineral exports (Bureau of Mines 1985a).

The statistics for the six-year period 1979–1984 illustrate clearly the decline of the domestic mining industry within the U.S. economy. Whereas in 1979 the value of domestic production was 1 percent of GNP, by 1984, at contemporary values, it had been reduced to 0.6 percent. Even more revealing has been the relationship between domestic production and imports. In 1979 the value of domestic production equaled that of all imports of nonfuel minerals at a figure of $24 billion; by 1984, however, the value of domestic production amounted to only 59 percent of the value of total imports (Bureau of Mines 1980, 1985a).

NONFUEL MINERALS IN THE U.S. ECONOMY

In 1984 processed minerals, domestic and imported, contributed just under 8 percent to the value of the GNP. Thus, while by no means dominant, the processed minerals sector can be seen as highly significant within the U.S. economy. When the actual role of this sector in the economy is examined, it can be appreciated that its importance extends well beyond its budgetary value. However, the sector depends crucially on imports to provide almost 22 percent of the value of processed materials. If current trends continue, dependence on foreign sources is likely to increase. The question arises whether the increased reliance of the United States on foreign sources of nonfuel minerals can be rectified by ensuring free trade on a worldwide scale, or whether it is definite proof of national vulnerability in a crucial area. If dependence on other countries is seen as vulnerability, then perhaps a greater degree of government control of trade is required. In any case, it could be argued that the idea of free trade does not apply to minerals, owing to the long lead times required at each stage, the capital intensiveness of projects, and the vertical integration on a global scale of many producers. Minerals are vital for achieving and maintaining general economic health and stability (Morgan 1985). They are essential for defense, agriculture, and the supply and conservation of energy. Minerals, and the technologies that they permit, are necessary for the maintenance of a sound environment; and, because minerals constitute a large component of international trade, they are significant in international relations. Minerals are thus basic to economic, social, and political life in a far more crucial way than can be illustrated by import statistics. Any vulnerability in this sphere is therefore of real significance to the nation. The per capita annual consumption of minerals in the United States is 41,000 pounds. Thus over 4 billion tons of new minerals are needed annually to sustain the U.S. economy.[1]

Nevertheless, the scale of the minerals market overall must be borne in mind. Within the U.S. economy, the annual value at the pithead or wellhead of coal, oil, and natural gas tends to be, in each case, greater than that of all minerals not directly used for energy-production purposes (Landsberg 1982).

The cost per ton of crude oil is usually some 20 times greater than the cost per ton of nonenergy minerals, so the effect of price changes in the oil economy is bound to be of an entirely different order from that in other sectors of the economy. The total export of metallic ores, concentrates, and scrap amounts to just over 1 percent of world exports, whereas the corresponding figure for fuels is in the region of 20 percent. However, substitution is often comparatively straightforward where fuels are concerned, and their scale of use is such that successful conservation programs can be implemented. In sharp contrast to fuels, minerals present a heterogeneous group with a wide-ranging variety of uses, which militates against easy substitution. Also, minerals are often employed on a scale that makes conservation particularly difficult.

Thus while nonfuel minerals are obviously of crucial importance as a fundamental component of lifestyle in the United States, they appear to be of only limited significance when viewed from the standpoint of world trade or the whole U.S. economy. The conclusion to be drawn from this observation must be that each mineral would require detailed investigation as to its critical and strategic nature. For each, there ought to be an appropriate policy that would in turn be an integral part of an aggregated national minerals policy.

THE U.S. STRATEGIC MINERALS BACKGROUND

Throughout history certain minerals have been strategic with regard to specific locations. The need for salt, for example, determined trade routes and locations for barter for many centuries. Seven thousand years ago, copper was the only strategic metal. It was vital for development in the Fertile Crescent (Mangone 1984).

During the last 70 years changes in the direction and dimension of the mineral trade have affected the United States. Before World War I the United States imported minerals on a modest scale; crucial among these imports was potash, of which Germany was the sole supplier (Alonzo 1980).

During the war, when potash supplies were cut off, the price rose sharply from $35 per ton to almost $500. This stimulated the development of an indigenous potash industry centered in New Mexico.

Imports of minerals began to increase markedly in the 1920s, but it was not until the 1970s that domestic production consistently fell below consumption. A. E. Eckes (1979) has traced in detail this growing import dependence of the United States. Charles K. Leith, George Otis Smith, and Josiah E. Spurr (three noted geologists) have stressed the vital relationship between mineral dependence and international relations.[2] Their work, during the interwar years, was aimed in part at alerting the policymakers and public alike to the increasing vulnerability of the United States in the sphere of domestic mineral production. World War I stimulated domestic production in the United States, but since that time the changing requirements of the economy as a whole and the vastly increased scale of consumption have increasingly militated against the achievement of self-sufficiency.

Undoubtedly, control over mineral resources did influence the outcome of the war. This has been documented in the case of the German war machine with special reference to the importance of the effect of nickel, copper, and tin shortages upon it. However, the United States and Britain also experienced supply disruptions that added greatly to the cost of the defense industrial base. The vital necessity of controlling sea lines of communication was obvious. The thinking on the question of minerals inherited from the experience of World War I highlighted the necessity of choosing between increasing high-cost domestic production and importing minerals on the one hand, and the need to construct a stockpile during peacetime on the other (Eckes 1979). As a consequence of the war, chromium, manganese, nickel, and tungsten could all be considered strategic. Then, as now, geopolitical risks were viewed with considerable concern; E. R. May (1973) states that substitution, protection for the domestic industry, and, as far as possible, the defense of foreign assets were all considered.

During the 1920s the elements of a minerals policy were discussed, including the relationship between controls and free trade. Underlying these discussions was an acknowledgment of the fact that no country could remain self-sufficient. This period also witnessed the imposition of tariffs by the United States government on imported ores such as bauxite, manganese, and tungsten and the establishment (with the collusion of the government) of cartels, particularly for copper. However, these measures were not successful. During the last half century, the United States has been opposed to cartels and other similar interferences with trade.

Mineral requirements exercised a strong influence on the conduct of World War II. The Germans in particular felt the need to increase supplies of iron ore, manganese, and molybdenum (Carroll 1968). Japan had access to far fewer minerals than it needed; it attempted to remedy the situation by developing mineral-rich colonies on mainland Asia. Toward the end of the 1930s the U.S. Planning Committee for Mineral Policy recommended that the government should implement a stockpile policy with the aim of ensuring supplies of manganese, chromium, nickel, mercury, mica, tungsten, and tin.

World War II witnessed a vast increase in both the volume and diversity of minerals required. Eckes' list of key strategic minerals (as of 1945) included manganese, vanadium, copper, zinc, lead, nickel, bauxite, chromium, cadmium, and tin. Eventually, after more than 20 years' discussion, the Stockpiling Act of 1946 was passed. While subsequent geopolitical disruptions have been limited, the wisdom of maintaining a stockpile has been demonstrated on a number of occasions.

In the immediate postwar years the United States was dependent on the Soviet Union for almost its entire supply of chromium and manganese. In 1948 the Soviet Union reduced shipments of these two ores. In 1976 the Angolan civil war led to a reduction in the supply of cobalt from Zaire; in 1978 civil war in Zaire itself resulted in a closure of the mines, at least for a short period. The price of cobalt increased from $6 per pound to approximately $40 (1978). However, so far the worst fears have not come true; indeed, during the Vietnam War, both copper and aluminum were sold from the stockpile.

It is clear that, since the pre–World War I period, the United States has become increasingly dependent on imports of minerals, and the list of minerals that must be classified as strategic has steadily increased. Furthermore, while the crises of 1973–74 and 1979 focused public attention on the problems of energy procurement, no event of comparable significance in the field of strategic minerals has yet taken place.

CRITICALITY

Defining what constitutes a defense industry gives rise to a number of problems. In particular, the production of weapons systems is so dependent on ancillary industries that it is difficult to demarcate a clear boundary for what actually constitutes defense. Furthermore, no detailed tracing of source routes to the ultimate minerals has been attempted even for the major strands of the industry. In broad terms, it has been noted that the Department of Defense is "heavily dependent on speciality metals and alloys" (Office of the Under Secretary of Defense for Research and Engineering 1981, 60).

Although specific data on the subject are scarce, some idea of relative mineral criticality can be obtained from an examination of the industrial base. The power exercised by a nation depends in large measure on its economic base rather than on any particular group of industries. In the case of the United States, the following four basic industries account for most of the mineral materials (Bureau of Mines 1983):

1. *Electrical Products*: tantalum, copper, platinum, tin, cobalt, aluminum
2. *Machinery*: tungsten, iron, chromium, zinc, manganese, copper
3. *Construction*: zinc, columbium[3], iron, manganese, aluminum, chromium
4. *Transportation*: titanium, lead, cobalt, platinum, iron, aluminum

Additionally, to ensure high agricultural productivity, fertilizers are a necessity. Essential to their manufacture are phosphates, nitrogen, and potash.

From the above industrial sector listing, it is possible to distinguish those basic industrial minerals such as iron, copper, and aluminum that are used in large quantities and are of low unit value. Of higher unit value and used in similar quantities are tin, zinc, and lead. Manganese is in a category of its own since it is on the one hand basic, being indispensable to steelmaking, but on the other it is of rather higher unit value and important in high-technology applications. A gradation can be discerned among the rest from chromium, used in comparatively large quantities of medium value and a high-technoogy material, to tantalum and platinum, used in small quantities and of relatively high unit value. In broad terms, a number of critical high-technology minerals can be identified for the key industries (Bureau of Mines 1983):

1. *Chemicals*: chromium, cobalt, titanium, platinum
2. *Electronics*: nickel, columbium, tantalum
3. *Energy*: columbium, titanium, nickel, cobalt, chromium
4. *Machinery*: cobalt, tantalum, titanium, tungsten, nickel
5. *Steel*: columbium, cobalt, chromium, nickel, tungsten, tantalum, titanium
6. *Transportation*: cobalt, chromium, nickel, titanium, columbium

All these minerals are of critical importance to the overall U.S. economic base. As the defense industry tends to become more and more high-technology oriented, fewer minerals are categorized as critical. Changes accompanying the rapid development of the electronics and electrical industries have necessitated systematic analyses of mineral requirements. Principal basic minerals can be listed (Chynoweth 1976) as follows: yttrium, indium, gallium, beryllium, germanium, tantalum, copper, platinum group, mercury, cobalt, gold, silver, tin, aluminum, tungsten, nickel, antimony, lead, rare earths, columbium, manganese, zinc, mica (sheet), and quartz (crystal). Important substitutes include bismuth, cadmium, silicon, boron, hafnium, and lithium.

As the sector of defense most totally dependent on high technology, the aerospace industry has been examined in a number of studies. Following an exploratory workshop, the needs of the U.S. aerospace industry were identified and those considered particularly critical were analyzed. Chromium, cobalt, tantalum, and titanium, together with alloys, were specially developed for rapid solidification technology and advanced aerospace composites were regarded as critical (Department of Commerce 1981). A study of the European NATO aerospace industry carried out for the Advisory Group for Aerospace Research and Development (Anderson 1984) concluded that, for various reasons,[4] the following critical minerals were most at risk in respect to their uninterrupted supply: chromium, cobalt, hafnium, manganese, columbium, titanium, tungsten, and vanadium.

A few studies have been made of individual weapons systems, the best known among them being that of the Pratt and Whitney F-100 Turbofan engine used in the F-15 and F-16 aircraft, for the manufacture of which the following strategic minerals are required (Siegel 1981):

1. titanium	:	5,479 pounds
2. nickel	:	4,597 pounds
3. chromium	:	1,537 pounds
4. cobalt	:	886 pounds
5. aluminum	:	715 pounds
6. columbium	:	163 pounds
7. tantalum	:	3 pounds

The M-1 tank engine requires approximately the following:[5]

1. chromium	:	343 pounds
2. columbium	:	25 pounds
3. cobalt	:	21 pounds
4. titanium	:	3 pounds

A similar assessment has been made of the MX system (The Raw Materials Group 1981; De Vore 1981):

1. aluminum	:	10,000 tons
2. beryllium	:	24 tons
3. chromium	:	2,500 tons
4. titanium	:	150 tons
5. steel	:	890,000 tons

Studies are currently being made of the mineral requirements for the Strategic Defense Initiative (SDI), which could include chemical lasers or x-ray lasers or neutral particle beams.

From these analyses, it is possible to identify minerals that can be appropriately designated as critical. However, the speed of change in high technology is such that levels of criticality are constantly changing. Through the Office of Industrial Base Assessment (OIBA) (Department of Defense), the data base of all the components required by the manufacturers of weapons systems is being assembled. Only when such an exercise has been extended to the raw materials at one end of the source route can a true assessment of current criticality be made. Clearly, there are as yet insufficient source route analyses to permit a definitive identification of minerals critical to the defense industries.

IMPORT DEPENDENCE

Although it is as yet impossible to identify conclusively those minerals that are currently critical, truly strategic minerals can be identified by examining statistics relating to domestic production, domestic consumption, and trade. The Bureau of Mines issues annual *Mineral Commodity Summaries*, giving data on the 85 to 90 important nonfuel minerals used in the United States. In 1984 the United States was import dependent for 52 out of a total of 88 such minerals. Dependence as expressed as a percentage of apparent consumption varied from 1 percent in the case of lime to 100 percent for ten nonfuel minerals. Those for which foreign reliance was complete were caesium (compounds), columbium, corundum, diamond (industrial stone), gem stones, mica (natural sheet), strontium, thallium, thorium, and yttrium. Among these, the only mineral showing a marked change over the previous eight years was thallium, for which import dependence rose from 9 percent in 1977 to 20 percent in 1978. Thallium is recovered as a byproduct from flue dust and residues resulting from the smelting of copper, zinc, and lead ores; although it was contained in ores mined in 1984, none was recovered.

In assessing vulnerability, various cutoff levels of net import reliance might be used. For convenience, 50 percent is a reasonable figure, although in relation to the general proportion of mineral consumption used for defense industries, it seems to be a rather high figure. The National Strategic Materials and Minerals Policy Advisory Committee (NSMMPAC) uses a vulnerability figure of 80 percent. If 50 percent is taken as the cutoff figure, the U.S. trade figures for 1984 indicate that 29 minerals were potentially strategic. This is higher than the corresponding number of 25 for 1979 and 22 for 1981. If 80 percent is taken as the cutoff figure, 17 minerals came under the definition of strategic minerals in 1984, more in line with the figures for the past eight years as a whole. While variations at the 50 percent level are partly due to changing import requirements, they also reflect the fact that statistics may often be either withheld or simply unavailable. Barite, previously listed as barium, entered the 50 percent vulnerability category in 1982. Import dependence of gallium has varied from 30 percent to 68 percent, but statistics have not been available for the past three years. Mercury has shown major changes, ranging from 26 percent to 64 percent, with 30 percent in 1983 and 60 percent in 1984. Similarly, silver, depending very much on market activity, has varied from 7 percent (1980) to 61 percent (1984). Changes among the vulnerable minerals recognized as such have been comparatively modest; for example, in the case of manganese the figure for import dependence has varied between 97 percent and 99 percent. The 1984 statistics also reveal that, usually for commercial reasons, the percentage of import dependence for 12 minerals was withheld. In a further five cases statistics were not available, and in one case (crushed stone) no figure was given. On the other hand, there was a net export of 18 minerals, including boron, lithium, and molybdenum, the last of these being considered strategic in NATO (Europe). The 1984 figure for the export of raw and processed nonfuel minerals was $20 billion.

In a series of regular unofficial publications, Bureau of Mines statistics are set out in the form of a bar diagram to illustrate U.S. net import reliance for selected nonfuel mineral materials (e.g., Morgan 1985). Recently 32 minerals have been included in the list. In 1984 the following minerals had a net import reliance exceeding 50 percent: columbium (100 percent), mica sheet (100 percent), strontium (100 percent), manganese (99 percent), bauxite and alumina (96 percent), cobalt (95 percent), tantalum (94 percent), platinum group (91 percent), chromium (82 percent), tin (79 percent), asbestos (75 percent), nickel (74 percent), potash (74 percent), tungsten (71 percent), zinc (67 percent), barite (64 percent), silver (61 percent), mercury (60 percent), cadmium (56 percent), antimony (54 percent) and selenium (51 percent). Interestingly, vanadium (usually included as a U.S. strategic mineral) shows an import dependence of 41 percent; in all but one of the previous eight years, this figure was considerably lower.

For various reasons, not shown on the bar diagrams, there are other minerals for which the United States is 100 percent import dependent. These are caesium, corundum, diamonds, thallium, thorium, and yttrium. The need for corundum has declined so steeply that it is no longer necessary to import it (as was indeed the case in 1984). Diamonds of course have a range of uses, few of them critical. For those that can be construed as critical, there are many ready substitutes. Both corundum and diamonds are obtained principally from South Africa, with Zaire being the second supplier of diamonds.

The production of both caesium and thallium is measured in pounds and therefore commercial stockpiles can be easily maintained. While thorium and yttrium production is in tons, it is of a lower order than that of the platinum group. All four have important critical uses: caesium in research and electronics, thorium in alloy production, thallium in electronics, and yttrium in communications. The United States is markedly import dependent for these four and they can indeed be considered critical minerals. Also, the United States is 91 percent import reliant for fluorspar, which is crucial in the production of aluminum and steel.

A further group of minerals to be considered is that for which statistics are not available. Critical among these are gallium, germanium, indium, and rubidium. They are of growing importance to the electronics industry. It can be reasonably conjectured that imports of germanium are comparatively modest (perhaps 25 percent of apparent consumption), but the United States relies on imports to a much greater extent for the other three. However, a new processing plant using raw materials is expected to produce both gallium and germanium, although without greatly affecting the degree of import reliance.

Finally, there is a comparatively large group of minerals for which data are withheld in order to avoid disclosure of proprietary information of companies. To this group belong antimony, arsenic, beryllium, bismuth, graphite (natural), hafnium, iodine, ilmenite, rutile, rhenium, tellurium, and zirconium. Using the last available figures and related data, it is possible to conjecture that import dependence is at least 50 percent for antimony, bismuth, graphite, iodine, rhenium, tellurium, and zirconium. It is over 75 percent for both ilmenite and rutile, the two sources of titanium, which is generally regarded as a strategic metal.

The more important of the two ores for the metal industry is rutile, and the most recent figure published by the Bureau of Mines (1985a) shows a 100 percent dependence. Thus to the original list of 22 critical minerals for which there is a high import reliance can be added a further 13, for various reasons omitted from the bar diagrams. The extent to which the United States is potentially vulnerable is reflected in this catalog of 35 minerals.

SOURCE VULNERABILITY

While any import dependence must represent a degree of vulnerability, the acceptable limit of such dependence is difficult to define. The pattern of world trade itself depends on comparative locational advantage. It is always possible that the movement of commodities may be subject to constraints of one kind or another. In attempting to structure a list of vulnerable minerals, therefore, it is necessary to examine the major sources and assess the probability of supply interruptions. As a first filter, the following factors would have to be considered in any such assessment:

1. potential political stability
2. distance and possible logistical problems
3. political orientation toward the United States
4. state of development of the mining industry, infrastructure, and so on
5. mining history

Having eliminated certain minerals from the list by adopting such a procedure, it should be possible to identify those that can be considered truly strategic. A second filter, to examine the various factors involved in detail, would follow the lines of the risk analysis model.[6]

A further question arises as to the percentage of mineral imports from a stable, reliable source that would render the United States invulnerable during a period of crisis. Since, in all cases, only a small percentage of end use (usually well under 20 percent) can be attributed directly to defense and related industries, a figure of over 50 percent would appear to provide a satisfactory margin. Furthermore, in times of tension, allocation would be imposed and minerals would be redirected from nonessential, nonstrategic end uses.

As a rough guide to the geopolitical aspect of U.S. import dependence, a gradation of external-source reliability can be established. Canada is obviously the most important and reliable supplier of minerals (Van Rensburg 1986), despite labor relations problems from time to time. Despite political problems, particularly those associated with its being a developing country, Mexico must also be considered a safe supplier because of its geographical position. Politically stable countries situated at varying distances from the United States, but relying upon sea lines of communication, though otherwise reliable, can present problems. Among these are the European Economic Community (EEC) countries, Australia, and Japan. South American and Caribbean countries, which have the advantage of propinquity, may be handicapped by sudden fluctuations in political stability. Potentially unstable less-developed countries and countries with centrally planned economies such as the Soviet Union and South Africa are the least reliable sources of import of strategic minerals. This is of course a very broad classification and within each level there are large differences. For example, whereas the Soviet Union has cut off supplies of strategic minerals to the United States, this has never been the case with South Africa. On the other hand, the current situation in South Africa, particularly when viewed in the context of American expectations, provides little scope for optimism.

Of the 21 nonfuel mineral materials included in the most recent Bureau of Mines bar diagram, the following eight are supplied principally by Canada: nickel, potash, tungsten, cadmium, zinc, asbestos, silver, and selenium. It should be noted that Canada is not the main supplier of any of the most vulnerable strategic minerals. Canada supplies more than 50 percent of potash, zinc, and asbestos. In the case of asbestos, however, although 95 percent of the imports originate in Canada, the remaining 5 percent from South Africa represent the only supplies of crocidolite and amosite (two important varieties of the mineral). If, for the remaining five minerals, the Canadian percentage is added to that from Australia, the EEC countries, and Mexico, a total figure of 50 percent net import reliance would result (except in the case of tungsten). The United States is 71 percent import reliant for tungsten of which Canada and Bolivia account for only 33 percent. There is also a marked reliance upon China.

Of the remaining 13 minerals, 95 percent of strontium imports originate in Mexico, while between them Japan and Spain supply well over 50 percent of mercury. While Australia, supplying 79 percent of imports, can be considered a safe source of alumina, the United States depends on Jamaica and Guinea to an equal extent for its bauxite requirements. In the case of tantalum, Canada is responsible for only about 8 percent of supplies, a large proportion of which come from Thailand (38 percent), Malaysia, and Brazil. Similarly, tin originates primarily in Southeast Asia with Thailand supplying 30 percent, Malaysia 23 percent, and Indonesia 16 percent. Barite supplies are dominated by China (43 percent), a large proportion of the remainder coming from South America. Mica sheet is supplied mainly by India (79 percent), but artificial mica has rendered this mineral less critical. Columbium is derived almost entirely from a single source, with Brazil dominating as the supplier of 75 percent of the total requirement.

The remaining five minerals include those normally considered most strategic. Suppliers of these originate primarily in Africa. South Africa is the major supplier of manganese, platinum-group metals, chromium, and antimony. In the case of manganese, South Africa and Gabon supply 60 percent of the ore (South Africa alone being responsible for 39 percent of ferro-manganese imports). Reliance on import from South African countries is even clearer in the case of chromium, of which South Africa supplies 55 percent and Zimbabwe 8 percent. In the case of platinum-group metals, South Africa supplies 49 percent, followed by Britain (15 percent).[7] It should be noted that the third most important source for both these minerals is the Soviet Union, which supplies 13 percent of the platinum group and 7 percent of chromium. Indeed, these are the only two important minerals for which the United States relies on the Soviet Union to any significant extent.

Antimony is somewhat different. South Africa supplies 40 percent of the oxide, while Bolivia is the second most important source of oxide and the main supplier of metal (together with ores and concentrates). China is also a significant supplier of metal (29 percent) and oxide (17 percent). Thus, as well as being a key supplier of barite and tungsten, China is an important source of antimony. Finally, the cobalt trade is dominated by Zaire and Zambia, which between them supply 51 percent of U.S. imports. Thus the key source of vulnerability seems to be southern Africa in general and South Africa in particular. Furthermore, 44 percent of vanadium also originates in South Africa. Clearly, columbium and mica sheet, both completely dominated by one source, represent extreme cases of concentration and would therefore require further investigation. Potentially unstable developing source countries are responsible for the major share of bauxite, tantalum, and tin imports, while China is the main source of barite.

Thus, from the first filtration, the following strategic minerals seem to require further geopolitical analysis: manganese, platinum group, chromium, antimony, columbium, mica sheet, bauxite, cobalt, tantalum, tin, tungsten, barite, and vanadium.

Of the minerals not on the bar diagram, but for which import dependence is 100 percent, the EEC countries supply over 70 percent of caesium, thorium, and thallium, while Australia provides 95 pecent of yttrium. Fluorspar may also be given closer consideration. Mexico provides 54 percent of the U.S. requirement with a further 28 percent deriving from South Africa.

Of the minerals for which statistics are not available, but for which there is a high import dependence, the entire indium requirement of the United States is supplied by EEC countries, while its rubidium requirement is wholly satisfied by imports from Canada. Of the minerals for which statistics are withheld, yet for which import reliance is high, Mexico supplies 61 percent of natural graphite, Japan 80 percent of iodine, Canada 44 percent and Hong Kong 20 percent of tellurium, and Australia, France, and Japan most of zirconium. Ores of titanium (usually considered strategic) are supplied principally by Australia, which provides 50 percent of the ilmenite and' 68 percent of the rutile. Since Canada provides a further 35 percent of the ilmenite, titanium ore does not seem particularly at risk. However, the production of titanium sponge from the ore is the more vital stage in the processing of the ore and this requires further study. The remaining minerals on this particular list are bismuth, of which Mexico supplies only 35 percent, and rhenium, of which West Germany supplies only 38 percent. In each case, the key sources of the mineral are in South America.

A first appraisal of U.S. strategic mineral vulnerability therefore indicates the need for a more detailed assessment of the following minerals and their geopolitical significance.

1. *Southern Africa*: manganese, platinum group, chromium, antimony, cobalt (vanadium)
2. *South and Southeast Asia*: mica sheet, tantalum, tin

3. *South America and the Caribbean*: columbium, bauxite, bismuth, rhenium

To these should be added fluorspar and titanium. Furthermore, it should be noted that seven of the minerals discussed (bismuth, gallium, germanium, indium, rhenium, tellerium, and thallium) are obtained as byproducts of base metal smelting (copper, lead, and zinc). With the decline of the U.S. metal industry, reliance on foreign sources for these minerals can be expected to intensify. Should the minerals policy of the United States government change, the outlook could be drastically altered.

DETAILED ASSESSMENT OF POTENTIAL STRATEGIC MINERALS

To identify the key strategic minerals, the factors already examined can be subjected to a more penetrating analysis. The investigation can also be advanced by a consideration of other factors and, in particular, substitution potential.

Antimony is used as an alloy, but its chief importance lies in its use as a flame retardant. For both these uses, and particularly the second, there are a number of substitutes. It is also relevant that the General Services Administration has been authorized to arrange the sale of antimony from U.S. stockpiles. Reserves are plentiful and the level of recycling is high (35–50 percent) (Crowson 1984).

Barite was included on the original list partly because of import dependence considerations, partly because almost 50 percent of the requirements were imported from China, and partly because it is one of the few nonfuel minerals for which the Soviet Union is markedly import dependent. However, while the use of barite in oil and gas well-drilling muds is vital, substitutes for it have been developed and, moreover, the industry is not directly defense related. Furthermore, although current domestic production is limited, the United States has the world's largest reserve base of barite.

Mica sheet is undoubtedly critical since virtually all of it is consumed in the electrical and electronics industries. There is no recycling; the United States is 100 percent import dependent and possesses only a very small reserve base. However, many substitutes have gained acceptance. These include reconstituted mica and other forms of built-up or bonded micas. Thus, although there are good grounds for considering antimony, barite, and mica sheet on a list of strategic minerals, in view of the ease with which they can be substituted they cannot really be classed as key strategics.

In the selection of strategic minerals, the import reliance of the United States, expressed as a percentage, was a crucial factor. Interpretation of the percentage presents problems. In some cases it is very specific and refers only to a specific year, while in others it has a more general application over a number of years. It provides an overall picture without giving any reason for the imports or the form in which they entered the country. In this analysis, an attempt is made to overcome the problem of specificity by examining import reliance over a period of six years.

Two elements related to the general nature of the statistics facilitate the identification of key strategics. These are the potential for conservation (and recycling in particular) and the scale of the U.S. reserve base.

The reserve base is that part of an identified resource that meets specified minimum physical and chemical criteria related to current mining and production practices, including those for grade, quality, thickness and depth. The reserve base is the in-place demonstrated resource from which reserves are estimated. It may encompass those parts of the resources that have a reasonable potential for becoming economically available within planning horizons beyond those that assume proven technology and current economics (Bureau of Mines 1985a).

Of the minerals already discussed, antimony enjoys a very high recovery rate, while the United States possesses approximately one quarter of the world barite reserve base. The annual production of rhenium is so small that sufficient quantitites of it could easily be stockpiled, and if necessary the mineral could be transported by air. While import dependence is high, the United States possesses 44 percent of the world rhenium reserve base.

Although bismuth can be substituted for its uses as a steel additive and in alloys, its major substitute in the steel industry would be tellurium, itself a marginal strategic. Recycling in the case of bismuth is insignificant and domestic production is at present low, but the United States does possess 7 percent of the world bismuth reserve base. Furthermore, there is a wide range of other potential sources of the mineral.

In contrast to bismuth (which is at best marginally critical), vanadium is extremely important as an alloying agent for iron and steel. Furthermore, in titanium alloys, crucial for the aerospace industry, vanadium cannot be substituted. In other alloys, vanadium substitutes are themselves strategic materials (e.g., columbium, molybdenum, manganese, titanium, and tungsten). Thus vanadium can be considered one of the most crucial minerals for the defense industry. However, U.S. import reliance for vanadium is currently only 41 percent, 5 percent higher than that for any year since 1977. Also, the United States possesses some 13 percent of the world vanadium reserve base, with domestic resources adequate for current needs. Thus the relatively high rate of imports is a response to price advantage rather than to domestic unavailability. Since over the past few years South Africa has been the major supplier (providing 44 percent of the imports), a certain piquancy is added to the situation. Using the criteria of defense end use and source instability, vanadium must be considered a key strategic. However, the comparatively low level of import reliance, together with the large reserve base in the United States, must result in its exclusion from the final list.

Titanium, which is highly critical and the main sources of which include South Africa, is in some respects similar to vanadium. Unlike other minerals, trade figures for titanium and its ores (ilmenite and rutile) are normally published. The main reason for this is that only minor quantities of the ores are used for the production of titanium sponge, the intermediate stage between ore and metal. It is difficult to obtain exact figures, but certainly well under 1 percent of the ilmenite and about 15 percent of the rutile produced are transformed into sponge as an end product. The figure for the combined ores is under 6 percent. However, the United States is highly import dependent for both ores. While up-to-date figures are unavailable, the most recent statistics put ilmenite import dependence at 74 percent (1982) and that of rutile at 100 percent (1979). The United States possesses approximately 10 percent of the world ilmenite reserve base, but only negligible quantities of rutile. Small amounts of titanium sponge were also imported by the United States, which has 25 percent of the total world production capacity.

In the absence of current data, it must be assumed that sources have not radically altered and that the chief supplier remains Australia (74 percent of ilmenite and 68 percent of rutile). The second major source of ilmenite was Canada (35 percent) and of rutile was Sierra Leone (16 percent). The dominant titanium sponge supplier was Japan (85 percent).

Since the defense use expressed as a percentage of total requirement of all first-order users was 44 (Congressional Research Service 1982, 11), it can safely be assumed that titanium is the most critical among the materials needed in large amounts. For the aircraft and space industries there is essentially no titanium substitute. Large-scale recycling is restricted almost exclusively to new scrap.

The question of whether titanium (and its ores) can be regarded as a strategic mineral thus presents complex problems. On the one hand, it is highly critical and there is a steady increase in demand; on the other, only a small fraction of the ore is converted into metal and the ore sources are not geopolitically vulnerable. It is therefore not really possible to classify titanium as a key strategic mineral.

Tin presents a different picture. Import reliance for tin at 79 percent has increased during recent years, but criticality is low. Tin has no unique properties, but an almost unrivaled combination of useful characteristics, many of which are of importance in the areas of electronics and transport. For certain alloys and for solder production there are no acceptable substitutes. Furthermore, there is no U.S. domestic production and the reserve base is negligible.

Geopolitically, tin is one of the most vulnerable of the materials used in comparatively large quantities. Its sources are concentrated almost exclusively in Southeast Asia (Thailand 30 percent, Malaysia 23 percent, and Indonesia 16 percent). Also, tin is one of the very few potentially strategic minerals of which there is a deficiency in both power blocs. However, during 1985 and 1986 the operation of the world tin market has floundered, with the result that there are at present large surpluses and the price is very low. Therefore, for reasons very different from those relevant to titanium, tin cannot be considered a key strategic mineral.

Bauxite, alumina, and fluorspar have several features in common. Import dependence runs at 96 percent for bauxite and 91 percent for fluorspar. In both cases, the U.S. reserve base is negligible. With 27 percent of capacity, the United States possesses the largest aluminum industry and recycling of the metal amounts to some 16 percent of apparent consumption. In line with this, the recycling of fluorspar in the aluminum industry is on the increase.

Since the defense use, expressed as a percentage of total requirement of all first-order users, was only 4 for aluminum, it cannot be considered highly critical. The same conclusion must be drawn for fluorspar, since its potential for criticality is related chiefly to the aluminum industry. Both have substitutes. Magnesium compounds and titanium may be used, at a cost, instead of aluminum. However, aluminum is likely to become increasingly critical as new alloys such as those based on ceramics are developed. Almost 80 percent of imported alumina comes from Australia. This does raise the question of the weight that should be given to the geopolitical factor.

The major supplier of refined bauxite is Jamaica, nearby and Western oriented, but this is closely followed by Guinea, a state noted for its mineral policies. Fifty-four percent of imported fluorspar originates in Mexico, but South Africa, which accounts for 28 percent, is easily the second largest supplier. Thus, in the case of both minerals, second suppliers are not as reliable as first suppliers. However, all things considered, and in particular the percentage of defense end use, neither can be considered key strategics.

Vanadium, titanium, tin, bauxite, and fluorspar fulfill, in different aspects, a number of criteria for being regarded as strategic. However, as their end uses, requirements, and supplies are subject to change, their status as strategic minerals should be kept under review.

Six of the remaining metals—tungsten, tantalum, columbium, manganese, cobalt, and chromium—are mainly used in the production of alloys, principally steel.

Platinum

Platinum, arguably the least important strategic metal, should be considered in the context of its metal group, which includes palladium, rhodium, ruthenium, iridium, and osmium. Within the pure concentrates in South Africa, platinum-group metals occur in the following percentage properties: platinum, 60; palladium, 27; ruthenium, 5; rhodium, 2.7; iridium, 0.7; and osmium, 0.06. The remaining 4 percent comprises gold (Anderson and Blake 1984, 12).

All the platinum-group metals are important in the electrical and electronic industries. Major applications in these spheres include the production of electrodes, relay contacts, thermocouples, and other such devices. While a far higher proportion is used generally in other industries, these are considered to be the strategic uses. While palladium and platinum account for a high proportion of these applications, rhodium, iridium, ruthenium, and osmium are used to a lesser degree. Of the last four metals, rhodium is particularly useful for fabricating items to be used at high temperatures; iridium is important for crucibles for melting nonmetallic substances at very high temperatures; ruthenium is used for high-quality resistors and (like iridium) as a hardener for platinum and palladium; osmium is used as an alloying agent to provide other precious metals with extreme hardness and corrosion resistance. A wide variety of alloys have also been developed within the platinum group itself and between its various constituents and other metals.

Thus, while platinum-group metals are important, their criticality is more difficult to assess. However their use for such items as jet engine glow plugs, special optical glasses, infrared reflecting surfaces, and fuel cell electrodes adequately demonstrates their potential. Their property of resistance to corrosion renders them vital for brazing alloys in those stainless steel systems needed to handle rocket fuels and oxidizers (Anderson and Blake 1984, 13).

In many of these applications, the platinum-group metals are virtually interchangeable. For general electrical and electronic applications, substitution is generally by other precious metals (gold and silver), but for resistors and thermocouples, tungsten is the main substitute. For the group as a whole, the proportion employed in defense-related industries, though small, is important and increasing rapidly. Nonetheless, it is the low level of criticality that places the strategic nature of the platinum-group metals in some doubt.

U.S. import reliance for platinum-group metals has been over 80 percent for the past five years. It has risen to 91 percent in 1986. Domestic production is negligible, although there is a small but significant reserve base. Indeed, there has been much discussion about mining the major deposit, the Stillwater Complex in Montana, and plans are in hand for an experimental project (1978). However, greater independence is achieved through recycling, the estimated capacity for which is 50–75 percent (Anderson and Blake 1984, 13). Platinum, for example, is virtually indestructible with a recovery rate of approximately 99 percent when used in the petroleum industry (in the cracking tower). However, in the electrical and electronics industries, the recovery rate is comparatively low.

The principal vulnerability of this group stems from geopolitical considerations. Almost 50 percent of platinum-group imports originate in South Africa, with a further 13 percent supplied by the Soviet Union. South Africa has estimated reserves of 30,200 metric tons (metal content) of platinum-group metals running to a depth of 600 meters. This constitutes 75 percent of the total world reserves. This proportion is likely to rise as mining takes place at greater depths. Meanwhile, the only alternative source is the Soviet Union, since the second largest source for the United States is listed in the figures as Britain. Britain, however, obtains all of its platinum-group metals from South Africa.

Thus, although the criticality of platinum-group metals may be in some doubt, there is a pronounced dependence on imports from source countries whose political disposition toward the United States cannot be guaranteed.

Each of the six alloying metals will be considered under the three analytic rubrics of critical uses,[8] import reliance,[9] and sources.[10] Their applications in the aerospace industry (the most highly sophisticated technological branch of the defense industry) are also included (Anderson 1984).

Tungsten

Tungsten has a range of unique properties. It is largely resistant to corrosion, imparts great hardness and strength, and is characterized by high resistance to wear. Its main critical uses are the production of tungsten carbides and alloy steels, which it endows with hardness at high temperatures. Tungsten carbides are vital when extreme wear resistance is required. Tungsten substitutes are (in all cases) inferior except for drilling bits and tools, which do not require extreme resistance. Tungsten is also vital in the production of superalloys and heavy-duty steels for armor plating. The aerospace industry is a medium-scale user of tungsten, although the need for tungsten superalloys in gas turbines and jet engines is of growing importance. In air-frame construction there are no alternatives.

For engines, tungsten is crucial, either by itself or with carbides, for tooling and in wear applications. It is also used to prevent wear erosion and oxidation in the form of plasma- or flame-spray. It is also of great importance in the manufacture of bearings and thrust rings. It is worth noting that some tungsten substitutes, for example titanium or tantalum carbides used for wear resistance or molybdenum used in tool steels and superalloys, are themselves strategic metals or compounds.

With one single exception, import reliance has been 50 percent or more over the past five years. In 1984 the figure was put at 71 percent. Recycling figures vary from 15 percent to about 30 percent, although this applies only to tungsten in alloy steels. (*N.B.* electrical uses allow no recovery.) While

U.S. consumption continues to rise and is expected to increase in the next five years at an estimated average annual rate of about 16 percent, domestic production is small and the reserve base is low. Indeed, a good reason for including tungsten among the key strategics lies in the widening of the gap between domestic production and expected consumption.

The major sources of tungsten are Canada (19 percent) and China (17 percent). China is expected to become an increasingly important source of tungsten imports. There is a fairly wide range of lesser sources, including Bolivia (14 percent), and this range may be contrasted strongly with that for the platinum-group metals. Thus, in view of the highly critical nature of the end uses and the increasing import reliance, tungsten deserves to be considered a key strategic.

Tantalum

Tantalum has a range of critical applications. It is used predominantly in the electronics industry, where the growth of tantalum capacitors for integrated circuits has been particularly marked. Its second major use is as a tantalum carbide for tool cutting edges. In the aerospace industry, tantalum is not used in air frames, but it is employed in engines, either by itself or in conjunction with columbium, as an alloy-strengthening element. It is invaluable for superalloys as a result of its high melting point and corrosion resistance. Most tantalum applications are substitutable, but with increased costs and often reduced effectiveness. Substitute hardening and corrosion-resisting agents are themselves generally strategic minerals such as columbium, tungsten, molybdenum, and titanium.

The major sources for import are Thailand (38 percent), Malaysia (10 percent), Brazil (10 percent), and Canada (8 percent). Therefore, almost 50 percent of supplies originate in Southeast Asia. It must be considered vulnerable in the medium if not the long term because it fulfills all three criteria, being highly critical and (apart from the small recycled supply) totally imported, largely from potentially unstable sources.

Columbium

Columbium is also important as an agent enhancing hardness and the resistance of alloys to corrosion. It increases heat and shock resistance, prevents internal corrosion, and acts as a stabilizer to prevent fatigue. Between 80 and 85 percent of the metal is used in the production of high-strength low-alloy (HSLA) steels, stainless, and alloy steels. It has a number of other critical but highly specialized uses in superconductors, ceramics, and fast nuclear reactor fuel cans. In the aerospace industry it is used to stabilize stainless steel for air frames and also for welding. In engine production it is employed chiefly to stabilize stainless steel and for precipitation hardening in nickel-based alloys and in wear coatings. Substitution is possible, but with penalties both financial and in performance. For HSLA steels, vanadium and molybdenum may be used; in stainless steel and high-strength steel and superalloys, tantalum and titanium provide substitutes.

The United States is 100 percent import reliant and the extent of recycling is insignificant. This reliance has been constant during the last five years, when domestic production has virtually ceased. The reserve base is also described as negligible.

By far the major source is Brazil, which supplies 75 percent of imports, followed by Canada (7 percent) and Thailand (6 percent). Bearing in mind that Brazil also possesses 88 percent of the world columbium reserve base, it can be appreciated that columbium represents the most extreme case of single-source concentration among key U.S. strategic minerals. Even though Brazil is geographically near and politically stable, effective reliance for supplies on three mines in a single complex at Araxa represents extreme vulnerability.

Manganese

Manganese is an essential ingredient for all alloy steels, over 90 percent of it being used in the metallurgical industry. Manganese is important for deoxidation and sulphur control besides its use in imparting hardness, toughness, durability, and resistance to abrasion. It is therefore crucial in the production of all-purpose steels as well as light alloys, and consequently is of great importance throughout the aerospace industry. Nickel may provide an expensive substitute, but desulphurizing agents are themselves scarce.

During the last five years the United States has been virtually 100 percent import dependent, with only 1 or 2 percent of its needs being met by recycling. At the same time, no domestic mine production of manganese has taken place in the United States and there is no reserve base.

The major sources of manganese ore are South Africa (31 percent) and Gabon (29 percent), Australia (17 percent), and Brazil (12 percent). Ferromanganese comes mainly from South Africa (39 percent) and France (30 percent). However, France is a secondary source because it depends on Gabon, South Africa, and Brazil for its supply of ores. Trade figures indicate that during the past five years the United States has become increasingly dependent on South Africa as a source of supply. Therefore, while not being a high-technology metal, manganese is nonetheless of crucial importance to various industries; and the United States is totally dependent on imports, primarily from potentially unstable sources. These factors are however somewhat offset by both the reasonably wide range of world producers and deep ocean mining, which will eventually alleviate shortages on a permanent basis. Until deep ocean mining becomes a reality, manganese must be viewed as a key strategic.

Cobalt

Cobalt is important for imparting strength and corrosion resistance to alloys at high temperatures. It is thus vital in the production of superalloys, which are required to function in increasingly hostile environments. Cobalt is also significant for magnetic steel production, but this use is declining. Within the aerospace industry, cobalt has wide-ranging significance, being used increasingly with nickel and chromium alloys for parts of the aeroframe while, at the same time, being of specific importance in engine construction. It is also vital for the production of machine tools and in a number of other crucial functions such as thrust rings, propellers, and engine bearings. Possible jet engine substitutes include alloys containing nickel, vanadium, chromium, and tungsten, but a number of costs, financial as well as performance related, are involved. Nickel can substitute for strengthening, but for wear resistance there is no satisfactory substitute. During the last five years the United States has been over 90 percent import reliant, the figure for 1984 being 95 percent. The remaining 5 percent is accounted for by recycling. Over the past five years U.S. domestic production has been nonexistent, although consideration has been given to using home sources. The United States possesses approximately 10 percent of the world cobalt reserve base, together with other subeconomic concentrations.

For imports, the United States is dependent on southern Africa, which provides over 50 percent (Zaire, 37 percent; Zambia, 12 percent). Furthermore, Zaire contains 25 percent of the world cobalt reserve base, the only other major reserves being located in China. Thus cobalt is highly critical and the United States is totally dependent on overseas supplies, primarily from the most volatile region of Africa. Furthermore, cobalt is the only concrete instance of a strategic mineral whose supply has been interrupted on a significant scale in the recent past.

Chromium

Chromium improves hardness, imparts high-temperature strength, and prevents rust in steels. It is therefore essential for stainless steels and, as a critical component, is irreplaceable at comparable costs. Approximately 70 percent of the metal is used in stainless steels and 25 percent in specialized alloy steels. Furthermore, chromium is the major component of the alloys in which it is found, normally comprising between 20 percent and 25 percent. For corrosion resistance, a level of below 12 percent results in markedly decreasing effectiveness. Of increasing importance in high technology is its use in the chromium-alumina treatment of the surface layer of alloys, imparting higher oxidation resistance. The aerospace industry is, therefore, one of the major users of chromium, particularly for the production of steel and nickel-based alloys. It is also vital for all heat-resistant stainless steels; there is no substitute for it in high-pressure, high-temperature corrosive environments. Such substitutes as may exist for its other uses are frequently very costly and less efficient than chromium.

Over the past five years U.S. import reliance has been high, varying between 76 percent and 91 percent, depending on the percentage recycled. During the same period there has been no domestic production and there is no U.S. reserve base.

Chromite and ferrochromium are obtained principally from South Africa (55 percent), Zimbabwe (8 percent), and the Soviet Union (7 percent). Furthermore, Zimbabwe and South Africa are the two countries with the largest chromium reserves. Of the two, South Africa holds approximately 85 percent. Thus chromium presents, in many ways, the most compelling combination of high criticality, import dependence, and source unpredictability. Furthermore, unlike manganese and cobalt, chromium will not be obtained by deep ocean mining.

Key strategic minerals have been identified on the basis of an analysis encompassing their critical uses, the reliance of the United States for their import from other parts of the world, and the location of their sources. According to some classifications, vanadium, titanium, tin, bauxite, and fluorspar would also be considered strategic. However, a strict application of the criteria would undoubtedly place tantalum, columbium, manganese, cobalt, and chromium among key strategics. Though at present slightly less at risk, tungsten is likely to become significantly more functionally critical over the next few years.

STRATEGIC MINERAL DEPENDENCE: SOME COMPARISONS

In sharp contrast to the United States, the Soviet Union enjoys a high degree of mineral self-sufficiency and is, for some, a major exporter. Indeed, it obtains nearly 40 percent of its hard-currency earnings from the sale of nonfuel minerals.

Soviet mineral policy objectives can be summarized under the following headings (Legvold 1983, 47–55):

1. the pursuit of mineral self-sufficiency
2. the development of the mineral trade as a means of expanding hard-currency earnings
3. the political use of minerals to integrate the economies of a number of other countries, including those of Eastern Europe, with that of the Soviet Union.

The Bureau of Mines bar diagram for the Soviet Union (1984) lists 33 nonfuel mineral materials, the same group as that discussed for the United States with the addition of molybdenum. The Soviet Union is import reliant for 11 of these minerals. In the case of five out of the 11, however, import dependence is 15 percent or below. The other six are: fluorspar (53 percent), barite (49 percent), bauxite and alumina (48 percent), cobalt (47 percent), tungsten (43 percent), and tin (33 percent). Since it has been demonstrated that the United States is vulnerable in respect to some six minerals, competition between the two superpowers for these scarce resources can be expected in the future. China is of course a major supplier of barite and tungsten, but the main potential theaters for competition seem to be Southeast Asia for tin and southern Africa for cobalt.

In examining the nonfuel mineral outlook for the Soviet Union (to 1990), J. S. Grichar, R. Levine, and L. Nahai (1981) have discussed projected growth and consumption statistics for various others. In each case, several projections are given. Adopting a median position, it appears that production is in phase with consumption for aluminum, cobalt (including supplies for Cuba), fluorspar, and manganese. Production of platinum-group metals will be well ahead of consumption by 1990, the only mineral seemingly at risk being chromium. On the other hand, in a detailed analysis, Levine computed demand for 1985 and 1990. He assessed correctly, that there would be no shortages by 1985 and considered that shortages for 1990 must remain conjectural. Similarly, W. Severin (1983) concludes that there is no real evidence that shortages will occur within the current decade.

Of particular interest, of course, is the mineral trade between the two superpowers. The Soviet Union is an important supplier only of platinum-group metals and chromium; in each case, it is well behind South Africa. It is in the platinum group that the Soviets have a major advantage. Whereas in South Africa the makeup of the platinum-group ores follows the ratio of platinum to palladium, 3:1, in the Soviet Union the proportion is reversed. Therefore, as the U.S. electronics industry increasingly adopts palladium as one of the major raw materials, it will become increasingly dependent on the Soviet Union. In the event of a boycott of South African supplies, the political implication of the trade relationship in this sphere between the United States and the Soviet Union could assume new significance.

The mineral-trading pattern of the Soviet Union has been variously interpreted, but account must be taken of the problems of climate, lack of infrastructure, paucity of capital investment, and a decline in labor force. Foreign investment has been encouraged, particularly from East European countries and Japan (Thomas 1986). Japan is much more dependent for mineral needs than either the United States or the EEC, if recycling is disregarded. There is a net import reliance for 18 of the 34 nonfuel mineral materials shown on the Bureau of Mines bar diagrams (the list for Japan and the EEC is the same as that for the Soviet Union with the addition of phosphate). Therefore, Japan, with extreme mineral dependence, but with a highly developed technology and, of course, capital for investment, would be well placed to enter into a symbiotic relationship with the Soviet Union. However, with the development of free-enterprise zones in China, Japan might diversify its external links in the sphere of strategics by forming closer ties with China.

For all the strategic minerals, the European members of NATO are approximately 90 percent import dependent, the figure varying from about 91 percent (cobalt and tantalum) to 100 percent (titanium) (Figure 2.1),[11] with the exception of

Figure 2.1
European NATO: Strategic Minerals Dependence

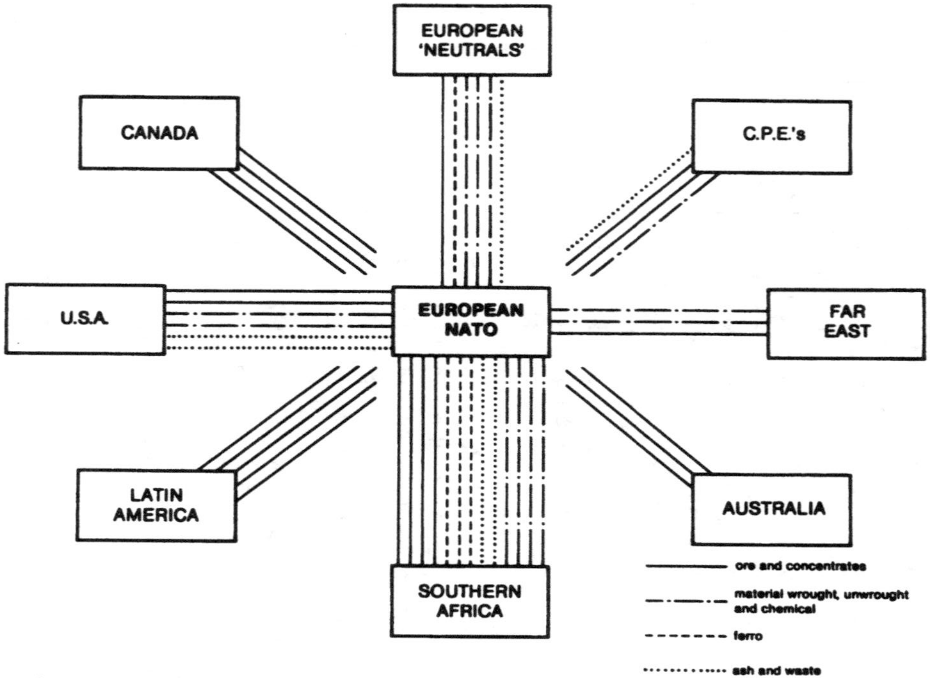

Source: Advisory Group for Aerospace Research and Development, North Atlantic Treaty Organization (AGARD/NATO), CP 356—Materials Substitution and Recyling in April 1984.

tungsten, for which the figure of dependence is 58 percent (Britain 56 percent). Furthermore, in the case of five minerals, France shows a degree of dependence below the average figures. None of the countries with a major defense industry has deposits on any scale, the only significant domestic production being of chromium (Turkey), ilmenite (Norway), and tungsten (Turkey and Portugal) (Anderson 1984); of these, domestic production of chromium in Turkey is the most important.

NOTES

1. B. G. Fourie address to the District of Columbia Bar, January 27, 1983.
2. W. C. J. Van Rensburg, author's personal communication, 1985.
3. Known generally outside the United States as niobium.
4. Principal among these were the potential geographical instability and concentration of sources, small-scale production, and technological and industrial problems.
5. Department of Defense 1984, internal paper.
6. The risk assessment developed for this study is discussed in Chapter 3.
7. British supplies originate in South Africa.
8. Hargreaves and Fromson (1983); Crowson (1984); Bureau of Mines (1985a).
9. Bureau of Mines (1985a).
10. Bureau of Mines (1985a); Anderson and Blake (1981).
11. Each line of the figure indicates a major source for one of the ten key strategic minerals (Anderson 1984).

3

Strategic Minerals: Sources

Key strategic minerals were identified using the following criteria: (a) criticality, (b) import dependence, (c) concentration of sources, and (d) technological, geopolitical, and other related factors. To keep the analysis within a manageable framework, it was completed as an elimination procedure in definite stages with only a restricted geopolitical input. The main problem arises from the geopolitical dimension, not only because it represents the least clearly defined aspect of the question, but also because it has to be evaluated on several different scales. With a view to achieving this end without sacrificing consistency, I have designed and tested a risk assessment model. However, since its application in each specific case would involve a detailed discussion, the preliminary selection of countries that constitute the sources of strategic minerals would need to be rigorous.

RISK ASSESSMENT

Risk assessment consists basically of two procedures: (a) dissection of the system under consideration into its key components, and (b) assigning to each the probability of a stated event occurring within a specific time period. The probability of such an occurrence expressed on a percentage scale or a numerical scale (usually zero to ten), is awarded either by the investigator or by adopting a variation on the

Delphi method. The probabilities as percentages or numbers for several related occurrences are then combined to produce a "magic" number that is effectively dimensionless and believed to encapsulate all the nuances of risk. In fact, with such a multivariate problem, the magic number defies interpretation since it results from a large range of inputs, any one of which may have dominated the calculation.

Producing one number has the advantage of simplicity and the apparently greater advantage of allowing comparability. However, if, for example, the risk to mineral supplies from Canada is given the magic number of one and that to supplies from Zaire eight, what exactly do these figures mean? For cobalt, the resource base of Zaire is greater than that of Canada for any individual mineral. It could be argued that Western industry suffered greater disruption from the Sudbury nickel miners' strike than from the insurrection in Shaba Province and the disruption of cobalt supply that ensued. Perhaps the government of Zaire is thought to be eight times as unstable as that of Canada. However, even if that were the case, there can be no direct correlation between governmental stability and interruptions to resource supplies. Furthermore, with its economy supported by International Monetary Fund (IMF) funding and its mining dominated by European expertise, the pro-Western character of Zaire may be less at risk than is commonly supposed. On the other hand, it is not beyond the bounds of possibility that a Canadian government could be elected on an antinuclear election platform or even one that aims at Canada's withdrawal from NATO.

Thus, given the conjectural nature of many of these probabilities, it is clear that a far more detailed "real world" appreciation is required before a really useful risk assessment can be made. A strong geographical element reflecting the real world is a sine qua non of risk assessment. The geography needs to be explicit and that is why the term "geopolitical" is preferable to "political." It is only by assessing the interplay of political, economic, and international factors at the specific location that the various risks identified can be monitored and even influenced by the foreign policy of the consumer country.

THE MODEL[1]

Geopolitical factors arise from the interplay of geography and politics on various scales. For convenience, the scales can be designated as local, regional, and global; but it should be realized that the exact definitions, especially of the first two, vary from one particular problem to another. In the present context, the term "local" is taken to mean "national," while the term "regional" refers either to areas encompassing contiguous states or to a region embracing different states, according to a generally accepted understanding of it. The scope of the term "global" is intercontinental and worldwide. For certain problems, one scale may be more important than another, whereas for others it may indeed be totally irrelevant. At the same time, the effect of a global political variable may well be felt on all three geographical scales; and, by the same token, a local geographical factor may influence political variables at all three levels.

The key geographical elements are location or place and (as a consequence) distance, both of which have to be considered on all three scales. Location implies the total geography, physical and human, but specifically in this case, it refers to scarce concentrations of strategic minerals. Distance varies according to the mode of transport and the route taken; but, on all three scales, there will be geographical influences. Depending on the scale of assessment, these may range from constraints at a particular port to problems of the regional infrastructure or global choke points.

The relevant political input could be described as policy, a resultant of such factors as ideology, type of government, and leadership. Examples on the global scale are world trading, defense policies of major powers, and the attitudes of multinational corporations. Regionally, policies may result from a common stance such as that adopted by most of the "frontline states" against the Republic of South Africa. Locally, policies are either national or within national subdivisions and concern such aspects as labor, tax, development, and environment. It is thus the interplay of decisions taken at different levels by the political actors, major, middle, and minor powers;

transnational and multinational bodies; and even individuals, interacting with the factors of location on different scales that produces the geopolitical environment (see Figure 3.1).

The full procedure for risk assessment of supply sources to the United States therefore begins with the establishment of the three criteria already discussed: (a) criticality, (b) import dependence, and (c) the concentration on limited sources. In applying the geopolitical part of the risk assessment model (Figure 3.2) that follows, geopolitical factors, global, regional, and local, which affect the production of the strategic mineral at its location, are examined. In turn, production of strategic minerals at a specific location exercises a geopolitical influence at the local, regional, and global levels. For example, superpower defense decisions, regional instability, and national mining policy are all factors that impinge on the production of a given mineral. Meanwhile, production may distort national economic development, affect the regional power distribution, and influence global diplomatic linkages. By separating the various activities on to different scales, the mechanisms involved can be clarified and more easily identified. However, this distinction is somewhat artificial because decisions made globally have a greater or lesser effect on regional and local geopolitics. For example, global strategic decisions affect world defense industries and generate demand for particular minerals. The interaction between these decisions and the geographical pattern of resources can result in changes in the infrastructure on all three scales and can exert influence on developments down to the level of a single mine.

The extreme case of dependence on virtually a single mining complex occurs in the case of columbium. The United States obtains 75 percent of its imports from Brazil and Brazilian production is limited solely to the mines of Araxa.

Similarly, local events are quite likely to exert an influence on the region and may even lead to global reaction. For example, industrial action or guerrilla activity may affect production at a given mine or in a mining region as a whole. The results will be felt regionally and if the material is considered sufficiently crucial, a superpower may physically intervene. Thus the scales are not discrete and the inputs from

Figure 3.1
Risk Assessment Model for Strategic Minerals

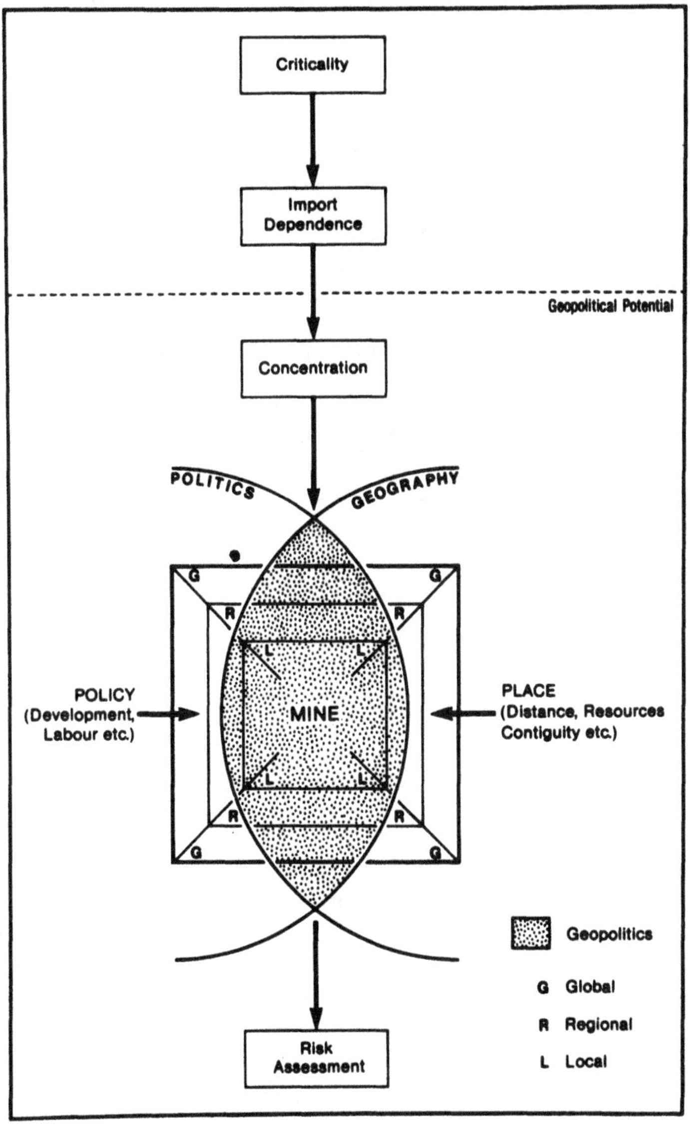

Figure 3.2
Geopolitical Risk Assessment General Model

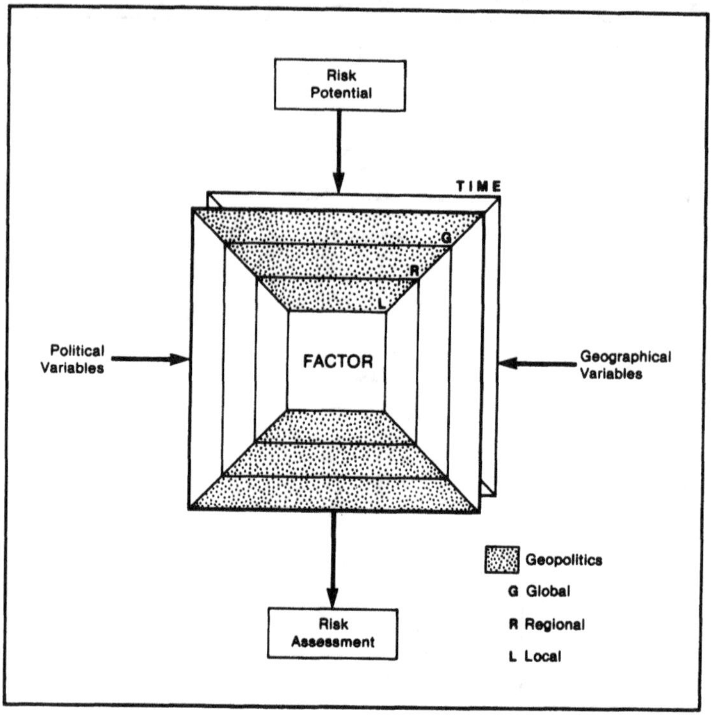

politics or geography on one scale may exercise marked influence on a different scale.

Furthermore, when the time scale is taken into account, the general tendency to regard long-term events as global and short-term events as local should be viewed with a certain degree of caution. Considerations bearing on NIEO or on changed supplier/consumer relationships may be global and long-term in character, but a price rise may be of a global and short-term character. However, in general, price changes result from temporary disruptions, market manipulation, or political instability, all of which act in the short term and may well be local. These are the most likely ways in which supplies may be disrupted and price is therefore a useful guide, since it encapsulates the range of influential factors: geological, technological, economic, and geopolitical. In the medium term, changing trade patterns, new sources of supply, and political alignment tend to operate on a larger scale, while in the long term, the reorganization of the world economic system or superpower military action have obvious global relevance.

The system is therefore very complex. It can be simplified objectively and rationally by the inputs of politics and geography on the various environmental scales. For each specific case under investigation, all inputs and all levels may not be relevant. By using the model, however, it should be possible to identify a series of key variables. From these, a realistic assessment of risk can be made. It may be possible to assign probabilities to particular events, but, in such a complicated situation, these must all be highly conjectural. It is obviously completely facile to assign numbers on a zero-to-ten scale for whole countries or whole industries. In order to produce a scenario of any value, the relevant factors require a detailed analysis; even such basic information as production statistics would need to be investigated.

THE MODEL AS APPLIED TO SOUTH AFRICA

The Republic of South Africa can be examined briefly as an example of how the geopolitical risks can be identified. The southernmost country on the African continent, the republic is in the last "ring-fence" position with a belt of independent African states to the north. Despite the Nkomati Agreement (1984) between South Africa and Mozambique, South Africa is faced with a whole series of problems, real and potential, along its northern border. Within the republic itself, the mining locations vary in vulnerability. A contrast could be drawn between the platinum and chromium mines of the northern Transvaal and the manganese quarries of the northern Cape (see Figures 3.3 and 3.4). South Africa's position is therefore globally and regionally isolated, thus giving rise to problems of

Figure 3.3
Platinum-Group Metals: Mines and Refineries

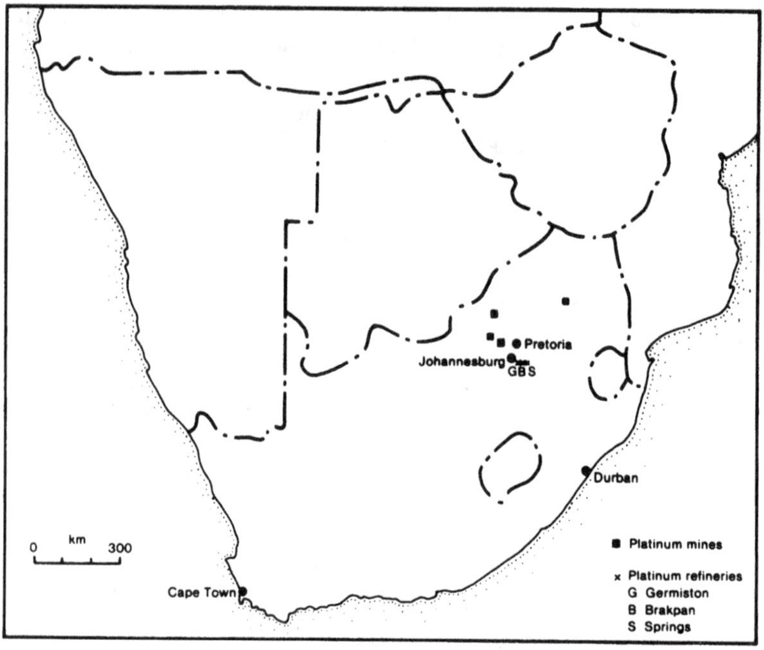

Source: Van Biljon W. J. A Review of the Geology and Major Economic Mineral Provinces of Southern Africa. Proc. Int. Mining and Metallurgical Congress. Johannesburg (May 1982).

maintaining not only the mining infrastructure (and the rail network of southern Africa in particular), but also the much discussed Cape Route (see Figure 3.5).

The geopolitical factors affecting mining (Anderson and Blake 1984) result, on the global scale, from the interaction of South Africa's possession of considerable and accessible deposits of certain key strategics with both the international opposition to its political regime and the arms buildup in general. Regionally, guerrilla activity in the "frontline states" endangers not only the infrastructure but also individual mines, particularly in the northeast. Locally, policies resulting from the racial composition of South Africa and the need to maintain a strong mining economy lead to many problems. Mining also exercises a profound influence on all three levels. Locally, it dominates the economy, producing some 50 percent of foreign-exchange earnings. Regionally, it has led to the development of a railway network upon which to a greater or lesser degree most countries of southern Africa depend, and also to the employment of expatriate labor on a large scale. Given the policies and attitudes of the government, it has resulted in South Africa becoming by far the most dominant regional power. Globally, the mining of key strategic minerals

Figure 3.4
Manganese and Chrome Mines

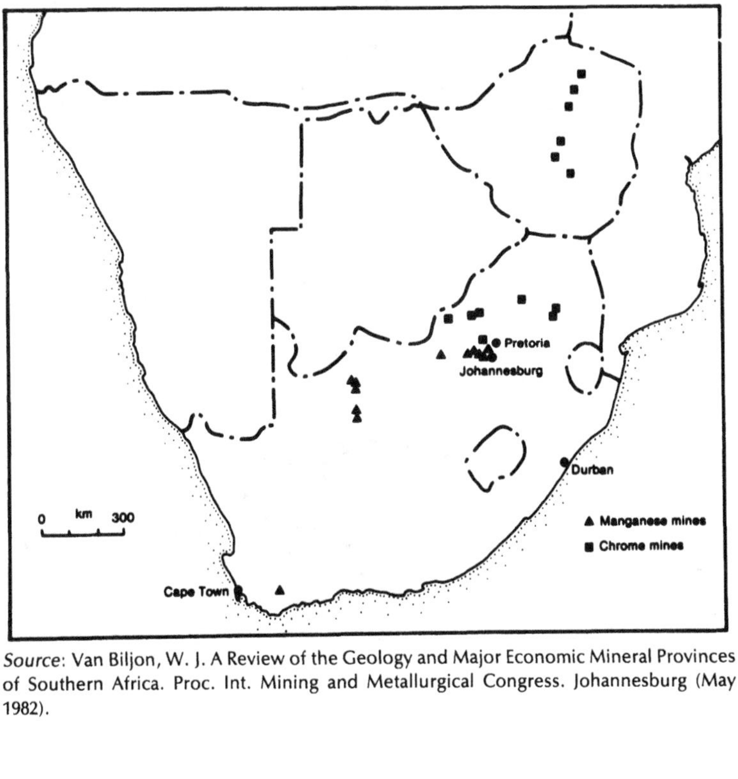

Source: Van Biljon, W. J. A Review of the Geology and Major Economic Mineral Provinces of Southern Africa. Proc. Int. Mining and Metallurgical Congress. Johannesburg (May 1982).

Figure 3.5
Southern Africa: Railways

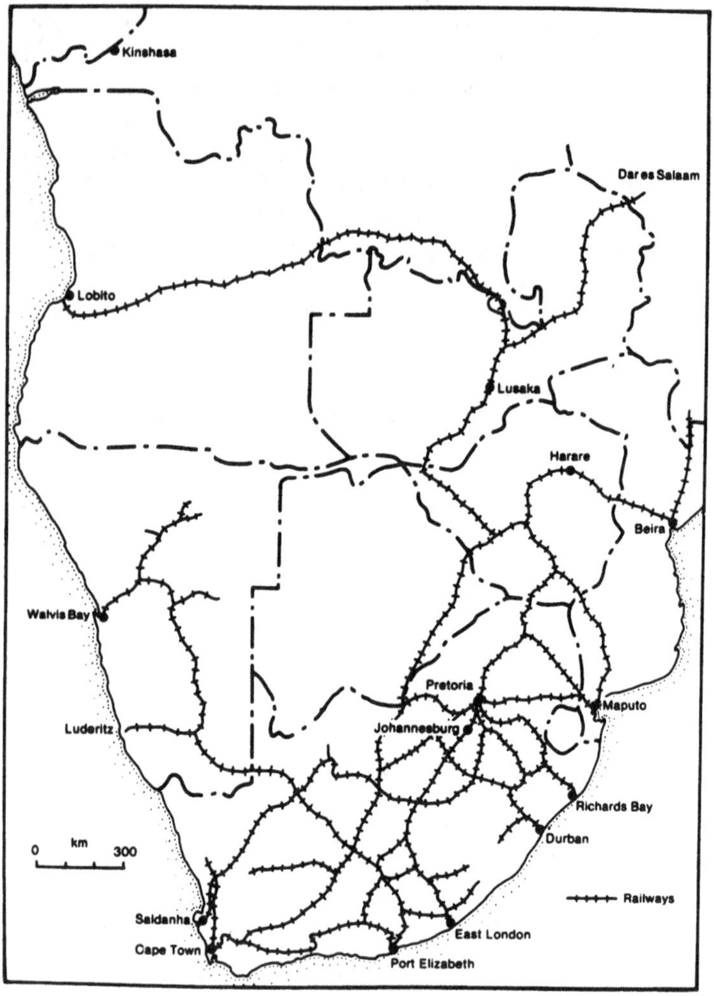

such as chromium, platinum, and manganese in South Africa has vastly influenced world diplomatic initiatives. Possible moves toward a rapprochement with NATO (or the West in general) result at least in part from the fact that South Africa is viewed as a vital mineral storehouse.

This selection of key factors involved in the functioning of such a complex structure is by no means complete; each factor could clearly be developed further. However, it does indicate that special consideration must be given in risk-analysis models to the local scale. Such a risk analysis would point to major threats to stability arising from internal causes. At the microlevel,[2] the port of Richard's Bay (the key exporting point for chrome ore and ferrochrome), for example, is particularly vulnerable. The assignment, therefore, of some arbitrary risk index to the country as a whole, or even to a particular native mineral, is clearly inadequate. Each mining area poses its own problems, and the vulnerability of specific routes and exit points would need detailed consideration. Furthermore, the situation in South Africa, as in other key supply areas, is fluid, and it is vital to monitor its changing policies and economic development.

If the main political input to risk assessment is described in general terms as "policy," then a wide range of actions on different scales, many highly technical, would be subsumed under that general heading. However, certain key policy approaches can be identified, some having a direct bearing on resource recovery, others resulting from problems connected with that recovery, while yet others may appear on both the input and output sides of the model. For example, competition between the NATO and Warsaw Pact countries for mineral resources has a bearing on the extraction industry but, basically, the factor of scarcity generates that competition.

In broad terms, however, consumer actions will affect the resource industry, while producer actions will influence the policy of consumers. Thus stockpiling and technological development, leading to improvements in substitution, recycling, and conservation, have a marked bearing on the quantity and quality of the mineral or minerals required. These are, however, often seen as countermeasures to the potential possessed by producers for using the materials as resource weapons. Thus even the potential for the formation of cartels or the emplacement of embargoes may be sufficient to produce policy reactions in the consumers. However, the distinction between producer and consumer may well be blurred as multinational corporations tend to straddle this boundary. In the development of many resources, they can influence both the input and the output sides of the risk assessment model.

KEY SOURCE IDENTIFICATION

In the previous analysis, the original list of minerals for which the United States is import dependent was reduced, using the criteria of: (a) obvious noncriticality, and (b) import dependence of less than 50 percent. The consequent elimination resulted in a group of 37 minerals that could be designated as strategic (including those that could only be regarded as marginally strategic). This list was then further refined by taking a number of factors (including approximate geopolitical factors) into consideration. Since the object of the present assessment is to identify what are geopolitically the key sources, this stage of analysis must be repeated in a more detailed and sophisticated form.

However, since the United States controls a worldwide trading network, the number of sources supplying the 37 minerals is of course extremely large. Therefore, to restrict selection to the key sources, only those supplying 20 percent or more of imports were included. In cases such as that of tungsten, for which the number of sources is diverse and none supplied 20 percent or more of the requirement, the leading two suppliers were included. The resulting list consisted of 18 countries together with Western Europe, which, because of the integrated nature of much of its mineral industry, was treated as a single unit.

Five major suppliers can be identified: Canada and Mexico, South Africa, Australia, and Western Europe. In 1984 Canada was the major source, supplying 11 minerals identified as being at least marginally critical. Only three of these, ilmenite, nickel, and tungsten, would be considered of front-rank importance in view of the fact that the others are partly or wholly substitutable. Mexico was a major source of five minerals (one of which was marginally strategic); Australia supplied six minerals (three of which could be considered strategic); among the six minerals originating in South Africa, five were strategically significant. Of the eight minerals from Western Europe, only one (ferromanganese, for which France is only a secondary source) could be identified as strategic. The prominence of South Africa as a major supplier stems mainly from the fact that the five minerals of which it is a leading supplier are all connected with high technology (mostly in the metallurgical industry), in addition to its growing political instability.

Of the remaining 14 sources, China supplied three minerals, Thailand two, and the remainder one each. By reassessing criticality, paying particular attention to possible substitution, it was possible to reduce this number to the following eight sources:

1. *Latin America*: Brazil and Jamaica
2. *Southeast and East Asia*: China, Malaysia, and Thailand
3. *Sub-Saharan Africa*: Gabon, Guinea, and Zaire.

This left two southern African countries, at present suppliers of minerals to the United States on a limited scale only, that could be regarded as vital. Both Zambia and Zimbabwe export a range of minerals, but, for this study, only those with a high degree of criticality were considered. In 1984 Zambia supplied 12 percent of U.S. cobalt imports and Zimbabwe 8 percent of chromium.

For each of the countries identified as a key U.S. source of nonfuel minerals, the risk assessment model was applied.[3] The procedure was restricted only to those minerals that are at least marginally strategic and only to aspects of the model relevant to the United States. This applied in particular to the feedback section of the assessment. For example, the effects of possessing a mine rich in a particular strategic mineral may well be vital for the economy of the country and may exercise a profound regional influence, but unless this has more than a general effect on the United States, it has been omitted. For the purpose of this study, the overall importance of the strategic minerals question should be such as to exercise some influence on the foreign policy of the United States.

CANADA

Canada is easily the most important source of U.S. minerals by volume of trade and the individual materials involved. Since it is a part of North America with the United States, the major forces involved at the regional level are those generated by its more powerful neighbor. Indeed, a major problem affecting all aspects of the national economy is Canada's lack of a separate identity. Bilateral trade between the United States and Canada is far greater than that between any other two countries, each being the other's primary source of mineral imports. For different reasons, both countries have developed anxieties about mineral dependence, the United States on imports and Canada on exports.

Canada is the leading exporter of minerals in the world and, for the United States, a key supplier of three strategic minerals: nickel, titanium (ilmenite), and tungsten. Furthermore, although Canada possesses in each case under 3 percent of the world reserves, significant amounts of cobalt, columbium, platinum group, and tantalum are exported to the United States. For the United States, Canada is the leading source of nickel and tungsten and the second main supplier of ilmenite. Tables 3.1, 3.2, and 3.3 show figures for: (a) imports as a percentage of total U.S. imports of each mineral, (b) percentage of world production for which the country concerned is responsible, and (c) percentage of the world reserve base in the country. In each case, the relative position of the country is given in brackets. Thus for these three strategic minerals, Canada is not only a major supplier, but it has the great advantage of possessing long-term reserves.

For its size, Canada has a well-developed transportation network with a marked east-west orientation. There are direct road and rail links with the United States, but trans-Canadian movement tends to be the dominant route. The network

Table 3.1
Strategic Minerals in Canada

Mineral	% of US imports	% of world production	% of world reserve base
Ilmenite	35 (2)	c.18 (2)	18 (2)
Nickel	38 (1)	19 (2)	13 (4)
Tungsten	19 (1)	7 (3)	19 (2)

comprises railways, road, pipelines, and waterways, with minerals in many cases providing the minimum guarantee needed to establish these facilities. In particular, northern extensions tend to be built primarily for mineral transportation, while class-2 railways are developed almost exclusively for mining. The Great Lakes–St. Lawrence Seaway (opened in 1959) is limited by the fact that it is only 27 feet deep and that the size of the lock is restricted. With the increase in the size of ships carrying minerals, these limitations have become important. However, the overall infrastructure of transportation is good and is under expansion.

Canada is also well endowed with energy supplies ranging from petroleum (onshore and offshore) to tar sands, uranium, nuclear energy sources, and hydroelectric power. Mining development is not likely to be affected by the problems of water supply. The mining labor force, comprising some 1 percent of the total labor force, is highly unionized, and large losses have been sustained through industrial action. This was particularly evident in the nickel mines at Sudbury. In general terms, however, the geographical infrastructure is well developed and poses comparatively few problems.

Yet the impact of government policies upon this has created a somewhat less favorable geopolitical environment. The level of direct U.S. investment is such that, particularly over the last ten years, Canadian anxiety over the maintenance of economic independence has steadily increased. This has been fueled by memories of American attempts to annex parts of Canada, and by such postwar developments as defense agreements as well as American cultural and media infiltration. Canadians now feel that they have been deprived in many sectors of domestic industry. In particular, the mining industry has always been heavily dependent on foreign investment, about 80 percent of which comes from the United States. In fact, 65 percent of Canada's largest corporations are either wholly owned or principally controlled by foreign corporations. As a result, two pieces of government legislation, the Foreign Investment Review Act (1973) and the National Energy Program (1980), have been introduced.

The basic objectives of this legislation are to ensure the continued development of the mining industry to Canada's benefit and to promote a policy of Canadianization (particularly of energy). This approach, along with the labor problems, has had a depressive effect on overseas investment. However, the diversification also promoted by this legislation has led to greater Japanese investment, particularly in the more favored joint ventures. Federal government interference is only one aspect of the mineral trade, since its operation depends also on the provincial governments, the mining corporations, the energy suppliers, and the consumers (both internal and external). Trade policy in general has been directed toward an increase in mineral processing, a diversification of secure markets, and improved competitiveness overall.

This political interest in mining is understandable when it is realized that minerals contribute some 9.5 percent of Canada's gross national product (GNP), over 30 percent of which is accounted for by nonfuel minerals. The provincial picture is, in many cases, far more compelling, with 85 percent of the GNP of the Yukon and Northwest Territories and 54 percent of the GNP of Alberta being directly attributable to minerals. Furthermore, minerals (fuel and nonfuel) account for between 28 and 30 percent of Canadian exports as a whole.

Thus, while Canada's geographical position is a strong positive factor, the political factor is complex and the consequences of the interplay between the two are difficult to predict. This is largely a result of the involvement of three key actors—the federal government, the provincial government, and the mining corporations (which may well be foreign owned). Government policies directed toward improving economic growth, sovereignty, and the quality of life generally, together with tax legislation, have resulted in strong federal involvement in the market and have adversely affected investment. For the United States, this government intrusion appears to be the principal threat. Any attempt to reorient the economy along lines of nationalism and increased competitiveness and to improve the structure of mining would inevitably weaken the U.S. position. In particular, Quebec, the second most important mineral province, with its separatist tendencies poses special problems. However, it must be concluded that through the innate natural strength of the economy and especially in the light of some amelioration of government policies, Canada remains the most secure supplier of U.S. strategic minerals.

On the local scale, the siting of mines gives rise to a further series of issues. In Canada, strategic vulnerability is usually a function of the geographical isolation of reserves, a good example of which is afforded by the location of tungsten deposits in the Northwest Territories. There are, however, exceptions. Titanium (ilmenite) and columbium are both mined in Quebec where the political situation has recently been sensitive. A completely different kind of vulnerability attends the supply of nickel. Although the principal mining sites in Ontario (particularly the Sudbury region) and Manitoba are secure, the bulk of the refining takes place in Clydach (South Wales) and Kristiansand (Norway). This transportational vulnerability is, however, peculiar to nickel with all other U.S. imports arriving directly from Canada by means of safe internal or coastal routes.

AUSTRALIA

Australia is in many ways similar to Canada as a major world mineral supplier, but it has its own brand of political problems. The key difference between the two countries lies in the extent of Australia's geographical separation from its major markets, particularly the United States.

Globally, this isolation represents the main threat to the country, with many of its near neighbors potentially unstable, a number possessing centrally planned economies. Australia is an outpost of the West that is particularly handicapped by a low population (15 million), with an area the size of the continental United States to defend. Therefore, its underpopulated (indeed, largely unpopulated) north and northwestern region is at strategic risk. With the very long sea lines involved, it would be difficult for the Western countries to coordinate aid to Australia.

On a regional level, this sense of isolation is considerably enhanced by Australia's distinctive peculiarity in an otherwise largely homogeneous part of the globe. The Philippines and New Zealand are the only other English-speaking countries and also the only other predominantly Christian countries. Apart from New Zealand, most of the nearby states are comparatively small with a high population density. Several possess mineral wealth, though none on the scale of Australia. Only in the cases of Malaysia, Indonesia, and Thailand is the economy as dependent on minerals as the Australian economy. Furthermore, any shipment from Australia is bound to pass through a number of choke points, the vital sea lines to Japan being particularly vulnerable to regional pressures of various kinds.

Like Canada, Australia is a major exporter of a wide variety of minerals, but as far as the United States is concerned three of these, bauxite, ilmenite, and rutile (both ores of titanium), are strategic (Table 3.2). Other minerals of strategic significance are asbestos, manganese, nickel, and tin.

As in Canada, the mineral industry is a major user of the transportation network. The remoteness of many mines from the populated regions of the country has led to infrastructural investment problems. In many cases foreign capital, with or without state assistance, has been employed in the construction of the railway network. In fact, the east coast aside, there is effectively no rail network. Rather there is a series of single lines to the coast from the various mining sites. While older

Table 3.2
Strategic Minerals in Australia

Mineral	% of US imports	% of world production	% of world reserve base
Bauxite (as alumina)	79 (1)	32 (1)	21 (2)
Ilmenite	58 (1)	c.26 (1)	6 (7)
Rutile	68 (1)	c.53 (1)	9 (3)

ports tend to be relatively small, the new ones, principally constructed for iron-ore export, can accommodate the largest ore carriers. With its vast reserves of coal, Australia has abundant energy resources. Although it is a net importer of oil, domestic energy resources are sufficient to meet the requirements of the mining industry. In contrast, water is scarce throughout the continent, except on its coastal fringes. However, due to the importance assigned to the developing mineral industry, a potentially serious water situation has been of less consequence than might otherwise have been the case. The Australian mining industry is more highly mechanized and therefore less labor intensive than its Canadian equivalent. Nevertheless, the Australian labor force is highly unionized and the industry is subject to occasional disruption due to industrial action. However, as in the case of Canada, geographical factors in Australia have combined to produce a very strong mineral economy. Since mineral production accounts for 10 percent of the GNP and minerals make up 32 percent of total exports, government interest in the sector is considerable. As in the case of Canada, the structure is complex, involving not only the central government but also the various state governments, the mining corporations, and the consumers, domestic and foreign. Furthermore, with 52 percent of the industry foreign owned and as much as 60 percent under foreign control, there have been similar apprehensions regarding the extent of foreign capital investment. During the 1970s the government introduced legislation to ensure Australian control of major investments in the future. However, although a large proportion of foreign investment originates in the United States, Japan rather than the United States is the principal market for the Australian mineral trade, unlike the situation in Canada.

The growth of Japanese industry, particularly its metallurgical sector, has generated the vast growth of Australian mining over the past 20 years. Rather than direct investment, the Japanese tend to favor long-term contracts. Long-term contracts are notoriously insensitive to price fluctuations. The Japanese prefer to import raw ores and concentrates, which are subject to refining and processing in Japan. However, it is worth noting that Japanese industry is changing (partly as a result of fuel costs) and is establishing plants offshore, particularly in North and South America. In the medium term, such a change of orientation in Japan's metallurgical industry could exercise a decisive impact on the Australian mining industry. If present trends continue, the Australian mining industry will be compelled to develop other bases. Indeed, Australia is at present looking to Southeast Asian markets in an effort to offset the potential decline in Japanese demand. However, in this region it encounters considerable competition from Canada, the United States, and South Africa for the export of coal, and from Brazil and India for iron-ore export. Thus in many ways the Australian mineral industry resembles closely that of Canada, the major differences stemming more from length of sea lines and position on the globe than from internal geographical factors. Furthermore, the two governments approach the various problems facing the mining industry in a similar manner. In both cases legislation has been introduced to safeguard the national interest, to improve the competitiveness of the industry, and to diversify the trade pattern.

At the local level the processing of bauxite, the key strategic, is divided between the Darling range in the south and the Gulf of Carpentaria in the far north. The latter location, which also includes a smelting facility, is all but indefensible. Ilmenite and rutile, together with zircon monazite, are mined from mineral sands around the core population regions of the east and west coasts.

SOUTH AFRICA

Were it not for the internal complications posed by apartheid, and its international ramifications, South Africa, the third greatest supplier of minerals in the world, would appear in many ways similar to Canada and Australia. From the global viewpoint, it contains a range of minerals vital to the West, but as in the case of Australia, sea lines are a source of problems. In fact, for various reasons, the Cape Route is considered one of the most vulnerable sea lines of communication. Situated at the midpoint of this route, between the South Atlantic and the Indian Oceans, South Africa is of great potential significance in view of the East–West confrontation.

On a regional scale, South Africa differs geopolitically from Canada and Australia in that it has land frontiers with less-developed countries. Moreover, the size of its armed forces and its probable possession of nuclear weaponry make it a relative superpower in the region. The "frontline states," while deploring the racial policies of South Africa, still maintain covert trading links and militarily are powerless to resist South African pressure. However, they represent a long-term threat to the South African regime, both by virtue of their very existence as well as their ability to provide bases for guerrilla initiatives. Nevertheless, there is no doubt that central and southern African countries depend heavily on the South African infrastructure in general and the rail network and the ports in particular. Exports from Zaire, Zambia, and Zimbabwe are generally dispatched from East London, although since the Nkomatic accord (1984), Maputo (Mozambique) is likely to remain open for such purposes. The other key influence on minerals is the need for labor, with repatriated wage earnings from the mines constituting a significant element in neighboring economies. Finally, the region is one of continuing turmoil, principally though perhaps not totally, generated by South African policies.

South Africa is a major world source of 17 minerals. It is no exaggeration to say that it is the main supplier of strategic minerals to the United States (Table 3.3). The four most commonly recognized key strategics, chromium, cobalt, manganese, and platinum-group metals (Office of Technology Assessment 1985), are to be found in South Africa. With by far the world's largest reserves of three of these key strategics to be found in South Africa (cobalt being the exception), their very location warrants such a classification.

Furthermore, in each case, in the event of South African supplies failing, the next largest (and in certain cases, the only other significant) producer is the Soviet Union. Indeed, for chromium and platinum-group metals, the Soviet Union appears on the trading list of the United States. Thus, even were South Africa as potentially stable as Australia or Canada, these minerals would be considered more strategic than anything supplied by those two countries.

Table 3.3
Strategic Minerals in South Africa

Mineral	% of US imports	% of world production	% of world reserve base
Chromium	55 (1)	27 (2)	84 (1)
Fluorspar	28 (2)	7 (5)	17 (1)
Manganese	31 (1)	11 (2)	71 (7)
(Ferro-manganese)	39 (1)		
Platinum	49 (1)	42 (2)	81 (1)
Vanadium	44 (1)	30 (2)	47 (1)

Table 3.4
World Production and World Reserves: South Africa and the Soviet Union

Percent of World Production				Percent of World Reserve Base		
Mineral	SA	SU	Total	SA	SU	Total
Chromium	27	29	56	84	2	86
Manganese	11	47	58	71	21	92
Platinum	42	54	96	81	17	98
Vanadium	30	33	63	47	25	72

South Africa is thus a supplier of more key strategics to the United States than any other country. Furthermore, for the four strategic minerals most dominated by South Africa, the Soviet Union is also a major producer, although it has comparatively limited reserves. Table 3.4 illustrates this point by listing the percentages of world production and of the world reserve base for both South Africa and the Soviet Union.

South Africa has a well-developed rail network linking mining centers with the major ports and also providing an outlet for countries to the north. At the same time, it has a series of magnificent deepwater ports, particularly the two newly developed bulk exporting ports of Saldanha and Richard's Bay. Richard's Bay, which is the major export point for chromite and ferrochrome, has some of the most sophisticated handling devices in the world. Minerals from South Africa's landlocked northern neighbors and Zaire are shipped from East London, a river port, while the manganese ore from the northern Cape is exported through Port Elizabeth. All such transportation to the West of course uses the vulnerable Cape Route. South Africa also has a well-developed road network that supplements the railways where necessary. With ample coal and uranium reserves, South Africa suffers no shortage of energy supplies for mining. However, the scarcity of water resources, particularly in the western part of South Africa, could well restrict future development. The labor problem arises from the fact that the vast majority of mine workers are black immigrants, either from neighboring states or from the various homeland regions within South Africa itself. Although wage differentials between whites and blacks have been improved (with the lesser skilled grades increasingly open to blacks), equality of opportunity is still a long way off. Thus, despite a certain degree of unionization, a good deal of labor unrest persists. While wages account for approximately 50 percent of mining costs,[4] one of the major benefits offered by the South African minerals industry has been comparatively low prices. Thus, although geographically the industry to some extent resembles those in Canada and Australia, the problems facing South Africa are potentially more severe.

Since the minerals industry accounts for some 20 percent of GNP and provides approximately 50 percent of foreign-exchange earnings, it is of great importance to the government. Although the state does operate directly in certain sectors (notably iron and zinc) and controls several others indirectly through the Industrial Development Corporation of South Africa, mining is generally left to private enterprise. Indeed, the mining laws and tax policies have been designed to encourage investment from abroad. Thus six multinational corporations account for most of the mining activity in South Africa, together constituting the Chamber of Mines of South Africa. Furthermore, through their provision of services various state organizations, such as the Geological Survey and the Council on Mineral Technology in particular, have put the domestic mining industry at the forefront of technological advances.

However, in any risk assessment of South African mineral supplies, race relations policies are probably of greater influence than those concerned specifically with mining. Since the sum of threats to South Africa, external and internal, is unique and exerts a crucial bearing upon the analysis, lengthier conclusions from field study are included here (Anderson and Blake 1984).

During the 1980s there would appear to be no serious external threats to strategic mineral procurement from South

Africa. International disunity and national self-interest have so far prevented any coordinated action against South Africa, and this situation is likely to continue for some time. However, extreme measures in South Africa itself might still provoke the international community to a more concerted effort against the present regime, though for the time being, internal threats seem to pose much the greater danger.

Despite its current depressed state, the mining industry has a great underlying strength, which results from:

1. the enormous scale of reserves
2. the highly developed infrastructure
3. the abundance of power for processing, benefication, and so on
4. the commercial stability
5. the technological expertise

Against this vast mining potential must be set the simmering resentment and frustration felt in particular by the black population. The key destabilizing factor appears to be the growth of an urban black population with a potential for conflict, ranging from civil disobedience to actual armed uprising. However, the South African security forces are so strong that it is doubtful whether internal unrest could, in the short term, become so serious as to affect the production of strategic minerals. It is more likely that the domestic populations of consumer countries will bring pressure to bear on their governments to discontinue the import of minerals from South Africa. If the situation in South Africa were to deteriorate to such an extent that a coordinated attack from the north occurred simultaneously with internal uprisings, the situation could well get rapidly out of control. In the background, the demographic composition of South Africa is rapidly changing due to the different birth rates of the black and white populations, changes in the physical environment, lack of water resources, and a steady encroachment of the desert.

Locally, the position of the Bushveld Igneous Complex renders the platinum and chromium mines particularly vulnerable. Not only are they near the major urban concentrations of South Africa where major confrontations are on the increase, but are also within easy striking distance of the "frontline state" borders. The northeastern border area is heavily forested and the potential for infiltration by guerrilla forces is high. Unlike these mines, the open-cast manganese quarries of the northern Cape, albeit close to the frontier, are in a semiarid environment with next to no cover at all. Thus, while the major mines are all situated near the frontier area, there are considerable differences between the west and the east.

MEXICO

Superficially, Mexico and Canada appear to share certain locational characteristics. Mexico, however, is distinguished by the fact that it is the only developing country in North America and as such has more in common with Latin and Central America. The United States takes a close interest in Mexico for a number of obvious reasons. During recent years, however, there has been an inclination by Mexico toward nonaligned nations, and, with Cuba and Nicaragua nearby, a question of supply vulnerability must remain. Nevertheless, it is difficult to believe that under existing political conditions Mexico could seriously consider a future for itself completely free of the United States.

Mexico's economy is potentially the strongest in the Caribbean, although its current international debt of approximately $60 billion (1985) has limited its scope for action. Furthermore, the Caribbean region has been politically unstable for well over a decade.

Mexico supplies fluorspar, bismuth, graphite, silver, and strontium to the United States, but of these, only fluorspar can be considered marginally strategic. Mexico is the source for 54 percent of the total import of fluorspar by the United States. It accounts for 14 percent of the total amount of fluorspar produced in the world (second largest producer) and has the fifth largest reserve base (9 percent). The major fluorspar mining occurs near the border with the United States.

Although the mines are not connected to the main rail network, they are situated in the most secure position within Mexico.

Mexico possesses a skeletal rail network with a line along the center of the country and one or two branches to key mining areas. However, a high proportion of the mining depends on road transport, although only 30 percent of the roads are paved. However, minerals are important in the transportation system, accounting for just under 30 percent of the total traffic. In 1980 the World Bank approved a loan to expand and develop its major rail network, while the government is at present restoring the cargo land bridge across the country from the Atlantic to the Pacific.

The transport network includes pipelines for petroleum and gas to the ports as well as directly into the United States. Mexico is the third largest oil producer in the West. It also has large gas reserves. Indeed, the energy balance is dominated by gas and petroleum, which account for 88 percent of primary requirements. A mere 9 percent of the energy consumed originates in coal, the remainder in hydroelectric power and the two nuclear power stations. However, there is a fast-growing demand for energy, which is bound to create bottlenecks in the future. Similarly, water resources, which are conspicuous by their extreme scarcity in the north, could present problems for Mexico's continued development. Labor relations have been reasonably calm, especially in comparison with a number of other Latin American countries. Mexico has experienced a major difficulty in maintaining a supply of skilled workers. Thus, although the geographical infrastructure reveals a number of problems relating to communication, energy, water, and labor, Mexico is comparatively well placed by the standards of other developing countries.

Minerals make up 74 percent of Mexican exports, of which nonfuel materials contribute 5 percent. Thus, with the industry highly critical for the economic development of the country, the government has been anxious to maintain foreign investment, mainly from the United States, Canada, and Japan. As a result of mining legislation (1975) and the policy of Mexicanization, the mineral sector consists of state-owned, partly state-owned, and private companies. Private mining companies contributed 49 percent of output and parastate companies 39 percent (1983). Almost 90 percent of all mining activity is conducted by the five largest mining companies and these tend to account for well over 50 percent of output. The government furthermore provides various subsidies and incentives for mining, its policy being to reduce the ownership role of the state.

Governmental stability has been maintained since the Mexican Revolution (1910), with power kept in the hands of a party that has consistently represented the country's major economic and political interests. During the 1970s there was a trend toward nonalignment, but in recent years the government has tended to move closer to the United States. The advantage of friendly governmental policy has reinforced that provided by geographical contiguity.

JAMAICA

Though more distant than Mexico, Jamaica is close enough to the United States to be geographically or locationally comparable to it. It has had a relatively stable government, particularly in the recent past. Jamaica is important to the United States for the supply of bauxite and alumina. Its reserves of both minerals are more abundant than those held by either of its neighbors, Guyana and Surinam. Averaged over a three-year period (1981–84), Jamaica has been a leading source of U.S. bauxite imports, providing 37 percent, and the second most important source of alumina with 11 percent. In 1984 it was the third largest bauxite producer in the world, accounting for 9 percent of the total world production. It possesses 9 percent of the world reserve base in bauxite, making it the fifth best endowed country. However, since 1985 the Jamaican bauxite industry has been in some difficulty.

Bauxite and alumina are transported by rail, road, and conveyor belt to the various processing plants and export docks. Exports have to pass through the narrow Windward Passage to the sea line to the United States. For its size, Jamaica has an efficient transport network. However, for energy it is almost totally dependent on imported oil, although the energy requirements of the bauxite industry are kept separate from the national grid. In view of the difficulties involved in oil supply, there are plans to use coal in the industry. Two major companies, Alcoa and Alcan, expect to complete conversion by 1988. Jamaica's water supply resources are sufficient to meet its needs, but provide virtually no opportunity for generating energy. Jamaica suffers from high unemployment, but labor unrest among highly unionized workers has not reached serious proportions. The main problem posed by the geographical infrastructure relates to energy resources and to reduction of costs.

Since bauxite and alumina account for over 70 percent of export earnings, Jamaica is beset with problems commonly experienced by single-commodity economies. Five major North American corporations, of which the government has acquired an interest in four, control the production of these minerals. However, government policy necessarily places a premium on bauxite and alumina. Though conveniently located near to the United States, any depression in the alumina industry will set off considerable reverberations.

BRAZIL

Brazil's geopolitical orientation differs considerably from that of both Jamaica and Mexico. While South America has long been linked with North America politically, economically, and strategically, recent changes in government in Argentina (1983), Brazil (1985), Peru (1985), and Uruguay (1986) from rule by the military to a popularly elected government, have tended to generate a fresh political orientation. There has been an increasing desire on the part of these countries to regard themselves as part of a South Atlantic and South Pacific land mass. Brazil and Argentina in particular have cultivated links with southern African countries. In the case of Brazil, of course, the potential for development is so great that the country may well become a major world power in the next century. It is certainly the major regional power in South America at the present time.

Brazil's vast land area has only been surveyed in a limited fashion and it would seem reasonable to suppose that major mineral deposits remain to be discovered. However, while the mineral sector has played an important role in the impressive economic growth of Brazil, it is not as significant as agriculture and manufacturing. Brazil possesses the world's largest reserves of at least three strategic minerals, of which columbium is the most important to the United States. It provides 75 percent of U.S. columbium imports and is the only relevant source of the mineral. Furthermore, in 1984 Brazil was the dominant producer of columbium (78 percent), and its reserves amount to 88 percent of the world base. Thus, despite Brazil's richness of mineral deposits, only one comparatively exotic metal possesses great strategic significance for the United States.

The transport network of Brazil is very thin and has been particularly developed to service the "iron quadrangle." A major line has also been constructed in the Carajas Province, but the columbium mines are served only by a recently constructed branch line and road. The mineral is produced from three open pit mines, while there is also a ferrocolumbium plant nearby at Araxa (Minas Gerais Province). Brazil uses a wide range of energy sources, but is highly dependent on imported oil (which accounts for over 40 percent of the energy balance). However, the development of domestic coal and hydroelectric power should alleviate the situation. Thus, while energy may be a limiting factor locally, the long-term supply of it poses no big problems. Water conservation measures in operation should ensure the supply required for mining purposes. Indeed, the columbium mining uses mainly recycled water, with additions from a nearby lake. Labor relations in the mining industry are very good, but the work force is comparatively small as a high degree of mechanization has been achieved. Brazil, therefore, appears to be taking its place as one of the major mineral source regions of the world, and, with a more developed transport network, there would appear to be few impediments to its exportation of strategic minerals to the United States.[5]

Government programs dominate the mining industry. A National Prospecting Program has been set up to oversee the development of the mineral industry. While state-owned companies account for about 70 percent of economic activity, government companies only account for 20 percent of total mineral output. The remainder is shared more or less equally between private Brazilian and foreign companies. Government policy can be described as supportive and is particularly active in metallurgical research and development of technology. However, mineral output accounts for only just over 2 percent of the GNP, and while future prospects are excellent, the economy is largely supported at present by other sectors. The mining industry in Brazil is in a strong position.

The fact that such a high percentage of columbium supplies to the United States have their origin in one small isolated area of Brazil must pose problems of vulnerability. In the total mineral portfolio of the United States, this is the most extreme case of concentration, since, even in the medium term, there appear to be no alternatives. Thus the effects of any constraint on supply, ranging from labor disruption to natural catastrophies, will be considerably magnified.

THAILAND

Thailand has long been a major world source of tin. In recent years, however, the mining of tantalum has also been developed. The strategic significance of Thailand for the United States as a source of tantalum cannot be exaggerated. Other minerals are also of importance, as well as petroleum and natural gas. Moreover, mining is the sector most likely to provide the impetus necessary for the expansion of the Thai economy. Thailand is the leading supplier of both tantalum (38 percent) and tin (30 percent) to the United States. Furthermore, it is the leading producer of tantalum (34 percent), possessing in addition the largest reserve base of the mineral (26 percent) among the noncentrally planned economies. Nonetheless, tin is the mainstay of the economy, despite the fact that the country is only the fifth largest world producer (10 percent). Tin reserves are comparatively widely spread and, with 9 percent of the world reserve base, Thailand is the second best endowed among tin-producing countries.

Though thin, the Thai railway network (which is mainly centered on Bangkok) does link up the main tin-mining areas of the peninsula. However, most of the remoter mining areas are totally dependent on roads, there being an attendant need for an improved infrastructure. There is sufficient energy for the development of mining, particularly given the scope for expansion of the petroleum industry. Hydroelectricity and thermal power from lignite are also important sources of energy with the former being capable of further short-term development. There are ample supplies of water for mining. Labor problems seem to be minimal. With improved infrastructure, the Thai mining industry could be rapidly developed. Thailand, unlike some of its neighbors, has been marked by political stability. Though the Thai government is interested in developing certain aspects of the mining industry, mineral production is still largely foreign owned. Since the economy is dependent on tin, representing 70 percent of the value of the total mining output, the government is anxious to encourage investment. However, such are the political uncertainties of the region as a whole, that little investment has been forthcoming.

It is useful to note that the further away a mine is from the capital the more open it is to the risk of being put out of operation. The tin mines are sufficiently important to the economic health of the nation that they are well guarded, but there are many in isolated locations.

MALAYSIA

Malaysia has a more balanced economy than Thailand and is the world's largest producer of tin, despite a recent decline in output. It also produces other nonfuel minerals, but petroleum and natural gas have been produced more than any other extractive products. Malaysia is second only to Thailand as tin exporter to the United States (providing 23 percent of imports). In 1984 it was the world's major producer (19 percent) with the largest reserve base (37 percent).

There is a restricted rail network serving the tin mines of western Malaysia. Both parts of the country are, of course, dominated by the sea, and shipping constitutes the basic mode of transportation of material. However, some of the major choke points of the world are located in the South China Sea, with an attendant vulnerability inevitable given the peculiarities of the region. Mining is powered by domestic oil, since there is little coal or developed hydroelectric power. Nevertheless, with Malaysia's burgeoning petroleum industry, energy should be abundantly and conveniently available. Furthermore, there is a plentiful supply of water and (despite some racial tension between the different ethnic groups) relatively few disturbances through industrial action. The transportation infrastructure of Malaysia, despite the problems of fragmentation and location, seems adequate.

Political stability, from the point of view of U.S. interests, has not been threatened so far. With an increasingly strong economy, based primarily on minerals, the government is encouraging investment. As in the case of Thailand, however, investors are not fully confident that the political situation will continue.

GABON

Gabon is a small state. Its economic development is directly related to its natural resources (particularly petroleum and, to a lesser degree, manganese). Gabon is also noted for uranium. Petroleum, manganese, and uranium together contribute approximately 70 percent of Gabon's GNP. Gabon is particularly important for the United States as an alternative source of manganese to South Africa. Indeed, it is the second most important supplier of manganese imports to the United States (29 percent). Gabon is the third most important world producer of manganese (accounting for 9 percent) and contains the fourth most important reserves (with 4 percent of the world reserve base).

A number of major roads constitute Gabon's transport network, with a rail system at present under construction. When the rail system is completed, manganese can be exported through the main port of Libreville. At present, manganese is transported by cable to Mbinda in the Congo, whence it is carried by the Congolese railway system to Pointe-Noire. This is undoubtedly one of the least satisfactory and most vulnerable networks presently used for the transport of strategic minerals. Energy presents no problem for mining, since there are large reserves of petroleum, natural gas, and uranium. Labor has been fairly quiescent in the face of pressures from an authoritarian government. As a small country with a limited but vital mineral industry, Gabon is well placed as a source of strategic minerals for the United States, particularly on completion of the railway link.

Government policy has combined nationalization with encouragement of investment. Since Gabon is rich in resources it enjoys the advantage of combining private with public enterprise.

Both the isolated position of the manganese mining near the frontier with the Congo and the extraordinarily tortuous communications lay the supplies open to attacks by the guerrilla forces active in the area from time to time.

ZAIRE

Zaire, a country rich in minerals, is best known as a major producer of cobalt. It is virtually landlocked, having a coastline only 37 kilometers long. It is surrounded by nine states, most of which have been politically less than stable: Congo, the Central African Republic, the Sudan, Uganda, Rwanda, Burundi, Tanzania, Zambia, and Angola. Since there is a recognized correlation between the incidence of conflict and the number of contiguous states, Zaire with nine different international borders must qualify as a potentially unstable country. Furthermore, almost all the mining in Zaire is located near the borders and is susceptible to disruption, given the activities of guerrilla forces.

Zaire, as the leading supplier of cobalt to the United States, provides 37 percent of imports. It produces 48 percent of the world output and holds 25 percent of the world reserve base, in both cases the highest percentage of any source country.

The railway infrastructure consists of four separate sections, two around the capital (Kinshasa) that are very short; a rather longer stretch in the north; and the main network that services the mining area of the southeast and Shaba Province in particular. There is no rail link from the mining regions to the coastline, nor, indeed, is there any port development on the coast. Three railways leave the country, one via Tanzania, one via Angola to Lobito, and one through Zambia and Zimbabwe to South Africa and the port of East London. As a result of management problems, disruption, and local conflict, only the last has provided a consistent outlet and thus the cobalt, together with copper, leaves from a South African port. During the 1978 crisis, cobalt was flown from Shaba, but now the price has declined so much that air transport is no longer economical.

Essentially, the road network provides feeder access to railheads and the river system. Though Zaire depends on imports for coal and ore, it has an extremely high potential for hydroelectric power development. Indeed, there is already in existence a network of high-voltage transmission lines from the Inga II Dam, which provides power for the Kolwezi copper cobalt refining and mining. There is no dearth of water in Zaire. The general unrest during the late 1970s was not directly due to labor problems. With an improved transport infrastructure, Zaire would be geographically well placed to develop its mining industry further.

After the traumatic political developments of the early 1960s, Zaire has been under a stable government for well over a decade. Despite interethnic conflicts and a secessionist movement in Shaba, Zaire has remained politically intact. Economic development has been slow. A very large debt has accumulated over the years, with Zaire increasingly dependent on the IMF. As in Gabon, the government encourages investment while retaining control over actual developments. Varying proportions of the major industries are under government ownership, and in the mining industry in particular the extent of government ownership has tended to increase.

Shaba Province, as the richest part of the country, is vulnerable to guerrilla attacks. Furthermore, with the existing political infrastructure, the Zairean government retains a rather tenuous hold on this outlying region. Its extreme importance to the West, however, offers some guarantee of a continued prosperity.

ZAMBIA

It has already been noted that neither Zambia nor Zimbabwe can be considered key sources for U.S. strategics at the present time. Nonetheless, they are in a volatile part of the world and do produce important minerals. Zambia is the second most important supplier of cobalt, providing 12 percent, and Zimbabwe is the second supplier of chromium with 8 percent. Copper and cobalt amount to almost 100 percent of Zambia's mineral exports, copper representing by far the larger proportion. The two minerals also provide the majority of foreign-exchange earnings and are therefore of critical importance to the economy as a whole.

The railway network consists of two lines, both of which are part of the railway line linking Zaire with South Africa. Sufficient energy for mining development is provided by hydroelectric power, coal, and imported petroleum. Approximately 75 percent of the total energy produced is consumed in the mining region. Water supplies are adequate, and labor unrest has been on only a limited scale.

As a result of the program of Zambianization instituted in 1966, foreigners constitute approximately 3 percent of the labor force. Zambia is geographically well placed to develop further its mineral industries. Mining is largely state owned or parastatal, although private enterprise is still given importance. Indeed, incentives are offered to foreign investors, with a favorable tax structure an additional attraction. Thus, at present, apart from the instability of many of its neighbors, the mining industry in Zambia is far less at risk than in most foreign countries. The cobalt mines in Zambia are, however, located near the Zaire frontier, an area considered the most vulnerable in the country.

ZIMBABWE

Zimbabwe, unlike Zambia, suffers from the geographical disadvantage of sharing a frontier with South Africa. Mining accounts for 7.5 percent of its GNP. It is a diverse and well-established sector. Production is comparatively small, but one or two minerals are of particular importance, notably chromium, even though chromite production has been halved during the past decade. With industrial development, most chromite is now exported in the form of ferrochromium.

The railway network consists of a central line with a series of side links. There are five exit points, the most important of which, economically, is that running to the port of East London in South Africa. All the main mining areas are linked to the rail network, with roads only important locally. Large supplies of domestic coal are available for energy production, but petroleum has to be imported. Hydroelectric power, however, constitutes the main source of energy. There is some potential for the development of hydroelectricity. Water supply does not pose serious problems. Labor problems are well controlled. Although representing a modest proportion of the economy as a whole, Zimbabwean mining is well structured with a good transport network. Given a stable environment, it has great potential for development.

The political structure of Zimbabwe is likely to change during the coming decade, with a change from a multi- to a one-party system presently under discussion. In the past, mining was privately owned, but the government has gradually extended its control over the industry, with results yet to be evaluated. The potential of the mining industry remains high, despite the political uncertainties both in Zimbabwe and in southern Africa as a whole. Chromium mines in Zimbabwe are located along a central belt in the middle of the country and are therefore in the least vulnerable position.

GUINEA

Guinea, the world's second most important bauxite producer, is located at the western end of the west coast of Africa. The region as a whole has been politically stable, although Guinea itself was until recently involved in the East-West ideological conflict. However, if by African standards the region is stable, on a world scale it still appears vulnerable to disturbance.

Guinea shares an inland border with Mali, one of the poorest countries in the world. It is extremely thinly populated. To the north are Senegal and Guinea-Bissau and to the south the comparatively stable states of Sierra Leone, Liberia, and the Ivory Coast. The principal disturbances in the region have thus been largely a result of Guinea's attempts to pursue a foreign policy independent of the French community.

Guinea is the second largest supplier of bauxite to the United States, providing 36 percent of the imports. It is also the world's second largest producer, accounting for 15 percent. It has, with 26 percent, the world's largest reserve base. The transport infrastructure consists of a skeletal rail system linking almost all the bauxite mining areas to the two ports, Conakry and Kamsar. Road transport is of local importance.

The only domestic source of energy is hydroelectric power, for which Guinea has a large and, as yet, unrealized potential. All other mineral fuels have to be imported. Where bauxite is extracted there is enough water, although the country becomes increasingly arid inland. With the departure of foreigners (even from skilled posts), the labor supply has become almost exclusively indigenous. Thus, for its limited industry, based almost exclusively on bauxite and its processed forms, Guinea is well served by its own skilled labor force. However, it has all the disadvantages of a single-commodity economy.

A centrally planned economy has been in operation since the inception of Guinea, although recently there has been something of a Westward orientation. Since minerals account for 95 percent of all export earnings (and bauxite, in some form, accounts for virtually all of this), the government has tried to encourage mining. There are three bauxite-producing operations in the country, one exclusively government owned and the other two privately owned, with a 49 percent government holding in each. The Soviet Union, the destination for a large proportion of bauxite exports, has invested in the industry. Guinea is thus at the heart of the conflict between the superpowers for resources. With the substantial reserves of bauxite available, geopolitical conflict over resources can only become more intense. The bauxite operations in Guinea are nearly all within a comparatively short distance of the coast and situated in the most densely populated areas. There is therefore no obvious source of local vulnerability.

CHINA

Although not yet a fully developed superpower (Anderson and Blake 1980), China is an important producer of minerals. Indeed, in time this importance is likely to increase significantly. China obviously has the potential to supply a wide range of strategic minerals to the United States, but at present it is important only for tungsten. Furthermore, tungsten imports into the United States come from a variety of sources, of which China is the second most important (supplying 17 percent of the total import). However, it is the world's major producer of tungsten (29 percent) and has the largest reserve base (36 percent). Clearly, the reorientation of its foreign policy, allowing for closer contact with the United States, is a geopolitical development of considerable significance. However, in any future relations between the two countries, the mineral component will be of comparatively small importance.

However, as its economy develops, China may forge closer links with both the United States and Japan in order to hasten its own technological progress.

Continuing supplies to the United States by China must depend on a number of factors; for example, development of the transport system, the requirements of other consumers and of the Chinese industry itself, and the continuing rapprochement between the two countries. However, given the rapidity of political change in China during the last two decades, accurate prediction of future developments is extremely difficult.

CONCLUSION

Any assessment of the relative importance to the United States of the key strategic mineral sources must be related to the current pattern of trade. There may be great potential in new places and significant changes may be forecast for the future, but the potential danger of immediate supply restrictions must be addressed.

By far the most vital major source country is South Africa. Geologically it possesses several of the most important strategic minerals, while technologically it has developed the most advanced mining industry in the world. Canada, Australia, and Mexico have potential for development, but, as yet, the range of minerals in each case is rather limited.

Of the other key sources, the geopolitically most crucial seems to be Brazil, since there are practically no alternative sources of columbium. Among the minerals required in larger quantities, one of the most important is cobalt. Zaire must therefore be considered another vital source. The other key supplier is Thailand, a producer of two strategic minerals and, at the present time, chiefly significant for tantalum. Two other countries of less importance now, but with obvious potential for the future, are Guinea as a supplier of bauxite and Gabon as a key source (after South Africa) of manganese.

NOTES

1. This model was developed and tested during work on a number of projects for the Ministry of Defence in 1984.

2. "Microlevel" here refers to the part of the port area in which strategic minerals are held.

3. For the analysis at the regional and particularly local level, where possible, personal experience was used. This was supplemented by data from a number of key sources, notably: Hargreaves and Fromson (1983), Paxton (1984), The Central Intelligence Agency (1984), Maull (1984), and most importantly, Van Rensburg (1986), together with the Bureau of Mines (1977, 1979a and b, 1980, 1981, 1983b, 1984, 1985a and b). The tables are based on those provided in *Commodity Summaries* (Bureau of Mines 1985a). It should be emphasized that the main element in any risk assessment must be the local factor, which can best be analyzed at the specific location.

4. This figure was calculated following interviews at the Minerals Bureau and Chamber of Mines (Johannesburg) and the Rand Afrikaans University at the University of Stellenbosch.

5. Under the Trade and Development Program of the International Development Cooperation Agency.

4

Strategic Minerals: Trade Routes

While the position of individual trade routes is generally dependent on the distribution of resources and the location of user industries, it is also influenced by economic considerations, as well as by such physical factors as meteorological conditions. The volume of trade flowing along any particular route is governed not only by market variables and economic growth rates, but also by other less tangible factors such as historical connections. Both the pattern and the volume of trade are subject to political and military constraints. While commodities in international trade may be transported by land or air, by far the greatest proportion (over 80 percent by volume) of trade is seaborne. Over 3,500 million metric tons of goods, consisting of half dry cargo and petroleum in roughly equal proportions, are moved by sea annually (Couper 1983). This is equivalent to 16,500 billion ton-miles of freight movement. Thus the effects of the dependence of the United States on import of strategic nonfuel minerals cannot be adequately considered without paying attention to both the stability of the suppliers and the security of the sea lines of communication.

The land lines bringing in trade (principally from Canada and Mexico) can be regarded as secure. Likewise, trade across the land frontiers of the United States by pipeline and across the Great Lakes is practically invulnerable to external danger.

Table 4.1
Strategic Minerals: Annual World Production (Comparative)

Rhenium	1	Columbium	1160*	Titanium	256000
Gallium	2			Chromium	680000
Indium	4	Cobalt	2240	Manganese	1840000
Germanium	8	Vanadium	2720		
Platinum	16	Tungsten	3760		
Beryllium	25				
Tantalum	26*				

* total for non-communist world only

Certain strategic minerals (the best known among them being the platinum group) are moved by air, a form of transport normally restricted to goods of comparatively low volume and high value. If key strategic minerals are ranked according to world production, the likely mode of transport can be identified. Taking rhenium, with an annual world production of about 12 metric tons as the unit (1), other examples are listed in Table 4.1 (as proportions of rhenium).

Depending on the circumstances, it would be possible to fly minerals listed in the first column. Those in the second and third columns would have to go by sea, the minerals in the second column by container vessel and those in the third by bulk carrier. In fact, titanium, depending on the particular route, could be transported either by container vessel or by bulk carrier. Thus, while for certain precious metals trade will be by air, the vast bulk of U.S. strategic nonfuel minerals, together with most petroleum imports, arrive by sea.

SEA LINES OF COMMUNICATION

The distribution of land and water on the surface of the globe is the controlling factor in determining the pattern of trade routes. There are essentially three major ocean basins, separated by north-south land barriers. The only potential for unimpeded east-west movement occurs at the margins of the polar oceans. Since the land stretches to higher latitudes in the north than in the south, the commercial scope of the Arctic Ocean is restricted by climatic factors. Therefore, the only major east-west routeway in use at present is along the fringe of the Antarctic Ocean. That the pattern of trade does not fully reflect this structure results of course from the fact that it is possible to leave the Indian Ocean through a number of straits along its eastern side, while interocean traffic can pass through the Suez and Panama canals. Sea lines of communication are thus focused on the two canals and the straits of Southeast Asia, together with the Cape of Good Hope and (to a much lesser extent) Cape Horn and southern Australia. These concentrations have become known as choke points and could be represented as the most vulnerable locations along the sea lines of communication (Anderson and Blake 1982).

The United States is the world's major trading nation. It therefore depends (directly or indirectly) on a number of the major sea lines of communication for its strategic mineral supplies. However, the number of routes is restricted for the major strategic minerals. For example, 75 to 80 percent of the United States' strategic mineral imports are transported through the waters of the Caribbean region (McMichael 1981) and would therefore have to pass through its various choke points. The infrastructure of the transportation route, with its attendant vulnerable choke points, can be discerned once the major world sources of key strategic minerals have been identified.

Columbium being transported along the route stretching from Brazil along the east coast of South America has to pass through Caribbean choke points. The route along the west coast of South America is chiefly important for tungsten and antimony, originating in Bolivia, and bismuth, originating in Peru. In addition to the Caribbean choke points, it is important to note the major constriction represented by the Panama Canal. Within the Caribbean region itself, alumina from Jamaica, passing through the Windward Passage, is the major mineral of strategic significance.

Tantalum and tin from Thailand and Malaysia are transported along the trans-Pacific route, almost certainly through the Strait of Malacca and undoubtedly through other Southeast Asian straits. The route by which tungsten from China and mercury from Japan arrive is of less strategic significance, though still involving passage through choke points around the Japanese islands. Sheet mica from India could of course travel east or west, according to circumstances. In either case, the route would pass through some of the world's major choke points, Malacca and Southwest Asia to the east and Bab el Mandeb, Suez, and Gibraltar to the west. Australia, a supplier of several strategic minerals, is a major source only of titanium ores and alumina. The trans-Pacific sea route from that continent passes through at least one choke point among the various islands of Southeast Asia.

For strategic minerals, the sea line from South Africa is by far the most important one. It forms part of the Cape Route. South Africa is the major source of manganese, chromium, antimony, and vanadium. It is also an important supplier of fluorspar. At the same time, cobalt from Zaire and Zambia and chromium from Zimbabwe are exported from South African ports. Even if Maputo (Mozambique) were to be used to transport chromium from Zimbabwe, the material would have to pass along the Cape Route through the choke point at the Cape of Good Hope. Further along the route, manganese from Gabon and bauxite from Guinea join the flow of strategic minerals through the choke point off the West African coast. The other trans-Atlantic route of significance stretches out from Western Europe. It is used for transporting not only mercury from Spain, but also a number of ores and semifinished strategics for which Western Europe constitutes a secondary source, most of the latter passing along the sea line through Dover Strait.

If the requirements of Japan and the European members of NATO for strategic minerals are added to those of the United States, the source areas already identified assume an even greater significance. The European strategic mineral supply lines are similar to those of the United States. They underline the dependence of the whole of the West upon southern Africa. Japan, on the other hand, has diversified its sources and is more dependent on centrally planned economies, though the routes involved are essentially similar to those listed.

Taking U.S. vulnerability with that of the West as a whole, the relative vulnerability of the different sea lines can be appreciated. Since part of the Cape Route between the Persian-Arabian Gulf and Gibraltar is important for the transport of both oil and strategic minerals for the United States, NATO members of Europe, and Japan, it must be considered the most significant strategic sea line. Other important sea lines involve passage through the various straits in the Caribbean area, Southeast Asian straits (in particular the Malacca Strait), the Strait of Bab el Mandeb, and Dover Strait. At the same time, the Suez and Panama canals impose great restraint on movement, being exposed to risk and easy to block or blockade.

CHOKE POINTS

Choke points (Anderson 1985a) occur where physical or economic sea-lane restrictions lead to a concentration of shipping within a limited area. Choke points occur most obviously where the land configuration produces narrow straits, although the true navigable width of the channel can only be ascertained from a bathymetric chart. Hazards to navigation in choke points may arise from a number of sources including, for example, other users, ferries, fishing boats, pleasure crafts, and wrecks, as well as islands and sandbanks. Choke points also tend to occur off promontories or at other places where trade routes naturally converge. Thus natural choke points can be seen off Cape Horn, the Cape of Good Hope, and Southeast Asia. As a secondary physical factor, meteorological and sea conditions may constrain the width of certain routes and thus produce high shipping densities.The economic necessity to take the shortest route wherever possible reinforces the tendency for choke points to develop off peninsulas where convergence occurs. Therefore, the location of choke points is usually determined by the configuration of the land, although this is by no means the sole factor involved.

Cartographic projections of shipping routes point to heavy concentrations of traffic, reflecting the most obvious choke points. Other choke points vary in shipping density according to time factors and, to some extent, chance. Choke points cannot be precisely defined in terms of specific numbers of ships within a finite area. Estimates of their number therefore vary widely from 100 to 200 (Department of State 1969; Smith 1974; Glassner 1978). If the most obvious configuration, the confined strait, is taken, the range can be narrowed down to between 109 and 119 choke points, each of a width of 24 nautical miles or less. Since 1982, following the general acceptance of a draft convention establishing the territorial limits at 12 nautical miles,[1] the use of such straits has necessarily entailed entry into territorial waters of another state. Thus 24 nautical miles is a significant measure, but the navigable width will almost certainly be much less.

The most important factor in defining choke points must be the ship count, although even this requires some modification, depending on the size and type of the ship and even the value of the merchandise carried. Since the value of the volume of trade carried per unit of time requires more sophisticated measurement, the usual variable listed under this heading is the number of ships. It is of course understood that these are warships or merchant ships of a reasonable size. By another direct method of evaluation, the cost of alternative routes, where possible, can be calculated. For example, rather than using Bab el Mandeb and the Suez a ship might take the Cape Route. There is, however, no alternative to the Strait of Hormuz.

While importance and vulnerability are obviously interrelated, the geographical characeristics of a choke point also affect its vulnerability. Of importance are the length and the most restricted width across the strait, the shallowest depth of water, and other characteristics of the configuration as a whole, both above and below the water. Vulnerability also depends crucially upon the political orientation of the riparian states. Dover Strait in this context obviously poses no problems, and Malacca at present few, but the Bab el Mandeb, which separates two opposing ideologies, is clearly at risk.

The most important choke points in the world are Hormuz and Bab el Mandeb, Dover Strait, the Dardanelles and the Bosphorus, the Straits of Malacca and Singapore, and the

Strait of Gibraltar. Of great significance, but less important, are Oresund, the Caribbean choke points (the Florida Strait, the Windward Passage, and the Mona Passage), as well as those of Southwest Asia (the Straits of Luzon, Sunda, Lombok, and San Bernardino). Though in many cases ship counts are not kept, it can be reasonably assumed that the following choke points, listed with their approximate annual total, generally record the largest number of ship transits: Dover, 350; Gibraltar, 200; Malacca, 140; Hormuz, 80; Dardanelles, 60; and Bab el Mandeb, 50 (Couper 1983).

THE CAPE ROUTE

From the viewpoints of the United States and its NATO allies in Europe, the Cape Route is uniquely important for a number of related reasons (Hanks 1981; Anderson and Blake 1982). Alone among the major sea routes, it hugs the coast for by far the greater portion of its length. Consequently, shipping along this route is vulnerable to interdiction by such hostile or potentially hostile littoral states as Iran, South Yemen, Mozambique, Madagascar, Angola, and Guinea. It is also the longest of the great sea routes. With the inclusion of the trans-Atlantic section, it is well in excess of 50 percent of the global equatorial circumference. Furthermore, the sea route passes through five well-defined choke points as well as being closely associated with Bab el Mandeb and Gibraltar. It is also, for some parts of its length, the major route for strategic minerals, while for others it is crucial for oil transportation, in particular to Japan and Western Europe (Figure 4.1).

In order to place the Cape Route in its geographical context, the great distances involved must be stressed. While the distance from New York to Southampton is approximately 3,100 miles, that from Gibraltar to the Persian–Arabian Gulf (i.e., the full extent of the Cape Route) is approximately 10,500 miles. Cape Town, the aproximate midpoint, is almost 7,000 miles from New York, over 5,000 miles from Gibraltar, and 3,800 miles from Rio de Janeiro. The distance eastward, from the Cape to the Strait of Hormuz, is 5,400 miles, only 200 miles less than the total distance across the Indian Ocean from Cape

Figure 4.1
The Cape Route and Choke Points

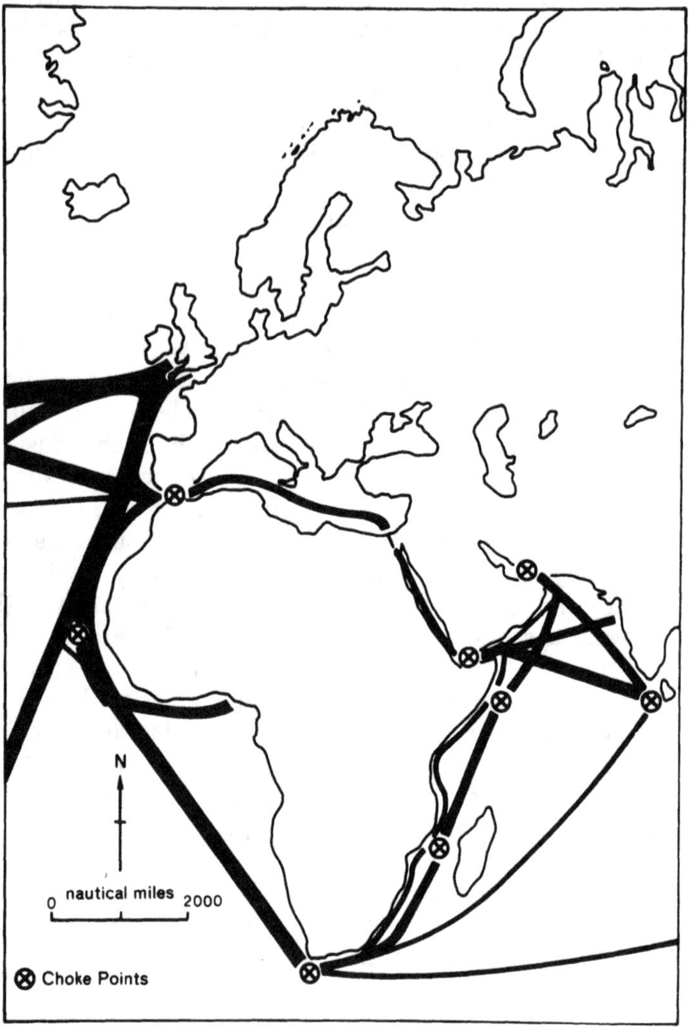

Town to Singapore. Thus to sail the full route at an economical speed would take more than a month for a normal merchant vessel.

Within the breadth of the route, the actual course selected will vary because of the need to maintain maximum economical speed and to avoid hazards. Economics dictate the shortest route, which consists of a series of straight sections as near to the coast as possible. The route around Africa runs fairly close to the coast for most of its length. This is particularly the case in the section to Cape Town from north of Durban, whence it leaves the coast for the only time before rejoining it at West Africa.

The first half of the Cape Route is susceptible to changes in climate and other physical conditions. With the onset of the Indian monsoon, the Somali Current undergoes a reversal that, during the summer (June to September) flows northeastward, reaching an average speed of seven knots. In order to avoid the problem of sailing against the summer current, vessels are forced to travel along the eastern edge of the route. During the rest of the year, the current is less pronounced, flowing southwestward as the gentle northeast monsoons drift. The tropical cyclones of the Indian Ocean, generated to the east of Madagascar, dictate the preferred course of shipping through the Mozambique Channel. To the south, off Durban, freak waves of up to 20 meters high, associated with sea and swell moving against the Agulhas Current, must be avoided by moving into deeper water away from the coast. During the past decade alone, eight serious encounters have been reported off this coast (Hydrographer of the Navy 1980). Further south, sailing close to the coast is unavoidable, both for economic and meteorological reasons. This area is prone to strong winds and heavy seas of the "Roaring Forties." In addition, the two sections that run most proximate to the coast—off South and West Africa—are vulnerable to frequent fogs.

A route's vulnerability is further increased by contrasts in the salinity and temperature of water masses that impede the detection of submarines. Marked differences in salinity do occur off the entrances to the Mediterranean Sea, the Gulf of Oman, and the Gulf of Aden. Of equal importance is the thermal layering that occurs over the Agulhas Bank off the Cape of Good Hope. This results from the mixing of the waters of the cold Benguela Current with those of the warm Agulhas Current. Therefore, at these critical positions, all of them near choke points, submarine operations are considerably facilitated.

Although the Mozambique Channel is a choke point, its use is determined largely by climatic factors and economics, as a possible alternative exists. Between the innermost Comoro Islands and the St. Lazarus Bank (over 90 nautical miles wide at its narrowest point), the channel still presents navigational problems. It is over 700 nautical miles in length, and the occurrence of various small island groups necessitates at least three major alterations of course. Passage through the channel is also affected by the Mozambique Current, setting approximately southwest, which can reach speeds of four knots.

At the Cape of Good Hope choke point, there are concentrations of shipping mainly for economic and climatic reasons. Several sea lines, including those from western Australia, the east coast of India, and the Malacca Strait, converge in the area between Durban and Cape Town. There is no separation of traffic, but ships are ordered to maintain set distances offshore.

Similarly, the sea line from the eastern seaboard of South America meets the Cape Route at the choke point off Guinea; and the Arabian Gulf, Red Sea, and trans-Indian Ocean routes converge at the choke point south of Socotra off northeast Africa.

With so many choke points and hostile riparian states, the Cape Route is the most vulnerable of all major sea lines. It could of course be argued that any interference with shipping would be more effective outside territorial waters, since it would involve no transgression of offshore boundaries and would be difficult to detect. However, any such advantages are likely to be counterbalanced by the difficulty of locating the target, even though both the aggressor and the victim could be identified by means of satellite imagery. At coastal concentrations, effective action against ships could be launched at a very low cost and without danger of detection.

Ships could be hit by local guerrillas in limited actions, which might well appear as accidents. Furthermore, were a hostile atmosphere to be created locally, insurance rates would probably be increased and owners and captains would be reluctant to allow their vessels to pass through the area. For example, while the Iran–Iraq War has not eliminated tanker trade from the northern part of the Persian–Arabian Gulf, it has severely affected insurance rates and has drastically altered the original shipping pattern.

The Cape Route is not only the most vulnerable, but also the most important sea line. However, no particular mineral follows the entire length of the route. The percentage of U.S. strategic mineral imports currently using part of the route is given in Table 4.2. The reliance of the European countries of NATO on imports follows a similar pattern. The percentages are in most cases high, while Japan is also served by a number of the same sources.

During the last 25 years, the Western alliance has ceased to dominate the region. There is no general treaty, Western bases are limited, and there is only restricted local support.

In marked contrast, the geopolitical strategy of the South Atlantic and particularly the Indian Ocean is thought by a

Table 4.2
Proportion of U.S. Strategic Mineral Imports Using the Cape Route

Strategic mineral	%	Strategic mineral		%
Aluminium	11	Manganese:	ore	60
Bauxite	36		ferro	39
Beryllium	5	Titanium:	ilmenite	6
Chromium	63		rutile	28
Cobalt	49	Vanadium		44
Fluorspar	28			

number of Western commentators to have been dominated by the Soviet Union, which has extended its influence, not only over the Rimland (Spykman 1944), but also beyond it into the Global Sea. The Heartland Power (Mackinder 1919) has thus broken through the classical constraints in some measure. A Soviet presence has been established in Luanda, regular visits are made by Soviet vessels to Conakry and Nacala, and a base has been constructed in Maputo where repair facilities have been installed. Both Angola and Mozambique, locations of strategic importance along the Cape Route, are indebted to the Soviet Union for arms supplies.

There has been a spectacular increase in the number of Soviet bases since 1968, particularly along the Indian Ocean section of the Cape Route, now extending from the Red Sea region (with the Dahlak Archipelago of Ethiopia) to Aden and Coetivy in the Seychelles Islands. At the same time, user privileges and the development of anchorages have extended Soviet influence northward to the Persian–Arabian Gulf and southward to Madagascar. Except in Berbera, there have been no real reversals in the Soviet Union's strategic influence. Centrally placed in this buildup is the South Yemeni island of Socotra, guarding the approaches to the Red Sea and astride the Cape Route. Despite having no natural harbors and being subject to the annual swing of the Indian monsoon winds, the island is important for its strategic position and affords a potential fallback position should political problems arise in South Yemen.

The Western response to the increase in Soviet influence in this area has been somewhat belated. Apart from establishing a base at Berbera, the West in general and the United States in particular have access to military facilities of various kinds in Oman and Bahrain in the Gulf region, at Mombasa and on the island of Mahé in the Seychelles.The United States also enjoys privileges in Mauritius and (with French cooperation) on the island of Mayotte in the Comoro Islands, as well as on Réunion and at Djibouti. However, of greater geopolitical importance is Diego Garcia, where a major base is being constructed under a combined British–United States initiative. It occupies a central place in the Indian Ocean region, at a distance of 2,000 miles from the Persian–Arabian Gulf and well removed from the Cape Route.

Given these changes in the balance of military power between the two blocs, the stance adopted by South Africa (with its modern bases at Simonstown and Durban and sophisticated communications center at Silvermine) would be a key element in any conflict. In the event of South African neutrality, major fleet maintenance could only be carried out in the Philippines, Western Europe, or perhaps South America, while ships proceeding from the South Atlantic to the Indian Ocean would need to refuel at sea.

Thus, more obviously than any other sea line of communication, the Cape Route provides an interesting illustration of the balance between politics, economics, and strategy. It is a route of vital importance to the West. However, it is uniquely vulnerable and links the world's major geopolitical flashpoints (i.e., the Persian–Arabian Gulf and southern Africa), situated in the two key areas from which strategic raw materials are exported to the United States.

KEY CHOKE POINTS

The Strait of Bab el Mandeb is 14 nautical miles wide, though divided by Perim Island into the Large Strait and the Small Strait. The Large Strait, nine nautical miles wide and 322 meters in depth, is used by shipping and is divided into separation lanes. The main navigation hazards include currents, winds, the tidal stream, and summer dust storms. However, the inherent political instability of the region poses a continuous threat to shipping because the territorial waters of four states mingle in the Strait of Bab el Mandeb: North Yemen, South Yemen, Ethiopia, and Djibouti. Of these, North Yemen and Djibouti are politically inclined toward the West, whereas South Yemen and Ethiopia are closely identified with the Soviet Union. With the opening of the Suez Canal, Bab el Mandeb became strategically important and, with the widening of the Suez, the significance of Bab el Mandeb has increased. Thus the strait is vital not only to Egypt, but also to Jordan, Sudan, Ethiopia, and North Yemen, for whom it provides the only maritime outlet, and to Israel for whom the port of Eilat is crucial. Furthermore, Saudi Arabia's major commercial ports are located at Yenbo (to which the trans-Arabian pipeline was connected in 1981) and Jeddah, both of which are located on the Red Sea.

The Strait of Gibraltar is approximately 36 nautical miles in length, varying in width from 12 nautical miles at its eastern end to 27 nautical miles at its western extremity, narrowing midway to a width of 8 nautical miles. Throughout the narrow section there are separation zones, but in contrast, for example, to the Strait of Dover, the Strait of Gibraltar is without radar surveillance. The average depth of the main channel is 360 meters. Apart from seasons of strong easterly winds, climatic conditions are generally good. As in the case of Bab el Mandeb, the importance of Gibraltar has increased along with the volume of traffic along the Suez Canal. Since so many littoral states depend on the Strait of Gibraltar for exit into the open ocean, the Rock of Gibraltar has long been considered a strategic base and has in fact been a British colony since 1713.

Dover Strait, which is the most important choke point in the world by ship count, has a minimum width of 18 nautical miles and an average depth of 20 meters. Climatic conditions are poor with frequent bad weather, including fog and storms, as well as a large tidal range. All these factors are likely to change rapidly according to time and location, as indeed are the many shoals and sandbanks. Furthermore, more than 200 wrecks are strewn over the area. The large volume of cross-channel traffic also poses a problem. Separation lanes allied to radar surveillance and other forms of monitoring have considerably reduced the risk of collisions.

The principal choke point in Southeast Asia, which is of worldwide importance by virtue of its geographical location, is the Strait of Malacca, which links the Indian and Pacific oceans. With the Singapore Strait, it has a total length of 600 nautical miles and its increasing importance has been linked to the increase in Suez Canal traffic and the development of the Japanese economy. The Strait of Malacca is approximately 140 nautical miles wide at its northern end and 8.4 nautical miles at the southern end, where the average depth is only 27 meters. Moreover, owing to the positions of islands and shoals in the south, large vessels are limited to a channel of only 2 nautical miles in width. Furthermore, climatic problems are posed by the varying and, at times, strong monsoon winds, migrating sandbanks, uncharted wrecks, and its daily civilian use by over half a million people who live along its shores.

The eastern continuation of the Strait of Malacca, the Singapore Strait, is 70 nautical miles long with a width varying from 15 to 2.5 nautical miles and an apparent depth of 25 meters. Navigation is further complicated by the necessity for a 90-degree change of course, shifting shoals, and occasional bouts of piracy. Alternatives to the Malacca Strait include the Sunda Strait, 12 nautical miles wide, 70 nautical miles long, and between 27 and 183 meters in depth. The preferred alternative is, however, the Lombok Strait, 12 nautical miles wide, but only 27 nautical miles in length and with a depth varying from 192 to 1,280 meters. The other two choke points of importance to the United States are the Luzon Strait (to the north of the Philippines) and the San Bernardino Strait, which is centrally placed in the island group. The former is 24 nautical miles wide and 5 nautical miles long with a depth varying from 55 to 183 meters, while the latter has a similar depth, but is only 2 nautical miles wide and 35 nautical miles long.

In contrast, the choke points of the Caribbean are of generally greater width than those of Southeast Asia. The Florida Channel is 82 nautical miles wide, 330 nautical miles long, and has an average depth of 1,042 meters. On either side of the Island of Hispaniola are the two vital choke points, the Windward Passage to the west and the Mona Passage to the east. The former is 45 nautical miles wide and 40 nautical miles long with an average depth of 396 meters, while the latter is 33 nautical miles wide and 50 nautical miles long, varying in depth between 61 and 274 meters.

The other most obvious and vital choke points (particularly to the United States) are the Panama and Suez Canals. Since its opening (1914), the Panama Canal has become absolutely crucial to the coastal shipping of both North and South

America. Moreover, it cuts down the sailing time for certain routes from the Far East to Europe. The Panama Canal is 43.5 nautical miles long and rises through a series of locks to 25.9 meters above sea level. Ship size is limited; maximum dimensions of ships are listed as 274.3 meters in length, 32.2 meters in breadth, and approximately 12 meters draught. The three sets of locks obviously delay passage. Transit time is usually eight to ten hours. Although an improvement program has been in operation since 1975, the key development envisaged is a new sea-level canal.The present canal tends to be somewhat congested with some 15,000 ships a year passing through it.

By contrast, the Suez Canal (opened in 1869) was constructed at sea level. Its length is 86.4 nautical miles and its original width was 21.9 meters and depth seven meters. It has been successively enlarged in 1956, 1961, and most recently in 1980. Its current depth is 19.5 meters. However, its width is going to be increased to 122 meters and its depth to 23.5 meters, allowing the passage of fully laden tankers of 250,000 deadweight tons. With a full passage taking about 24 hours, approximately 25,000 ships transit the Suez Canal each year. However, as was clearly demonstrated during the period 1967–1975, it is easy to block the canal. The Cape sea route will, therefore, always retain a strategic significance.

U.S. MERCHANT MARINE

We have discussed the U.S. dependence for strategic minerals on the outside world, described the major sources of supply, and pointed to the variety of technical and political stresses affecting the infrastructure that connects suppliers with consumers. It should be added that the United States is becoming more and more dependent on overseas markets and resources and, therefore, on the sea (Hayward 1979). It is obvious, given their relative geographical positions and trading patterns, that the Western countries are far more dependent on seaborne trade than Eastern countries. Indeed, the U.S. annual volume of such trade is approximately 875 million metric tons, with that of the Soviet Union a mere 202 million metric tons.

For NATO members as a whole, the average annual trade per country is over 200 million metric tons. Regarding aggregate international trade (exports and imports), the centrally planned economies are responsible for 4 percent of the crude oil, 7 percent of the oil products, and 8 percent of the dry cargo of the aggregate figures, whereas for the market economies, the respective figures are 40 percent, 50 percent, and 67 percent (Couper 1983).

The world fleet consists of just over 76,000 vessels, with a total cargo capacity of 422.5 million gross metric tons (Lloyds Register of Shipping 1985). The Soviet Union (with a total capacity of 2.45 million gross metric tons) ranks fifth in the world, with the United States as the sixth (19.3 million metric tons). Thus, despite differences in trading requirements, the Soviet Union owns a larger proportion of the fleet (5.9 percent of the total tonnage) than the United States (4.6 percent). However, from a strategic viewpoint, the actual number of ships is also of significance (especially in the case of the United States, which uses a wide range of trading routes). In this, both superpowers appear to be particularly well placed. Japan has the largest fleet (over 10,400 ships), but the Soviet Union is second (7,095), and the United States third (6,441). For both the superpowers, the average tonnage is comparatively low, being approximately 75 percent of Japan's and only 10 percent of Liberia's. Again, it ought to be emphasized that, with the Soviet Union dependent predominantly on land routes for trade and the United States on sea routes, there is a paradox in the figures for the percentage of world fleet owned by the two countries (9.3 for the Soviet Union and 8.5 for the United States).

Aggregate numbers and tonnages represent only a gross measure of any merchant fleet. Of greater relevance are the number and carrying capacity of the major elements of the fleet.

The United States is import dependent for a range of strategic minerals. Although a major proportion of these are transported in general cargo ships, a few (such as manganese) are required on a large enough scale to need ore and bulk carriers. Increasingly, as the system becomes more sophisticated, container ships will play a key role. The United States owns 136 ore and bulk carriers, representing 28 percent of the ships in the world fleet and 1.8 percent of the tonnage. The Soviet Union's cargo fleet is of similar proportions, numbering 141 carriers, representing 2.9 percent of the world fleet and 1.9 percent of the tonnage.

Bearing in mind their respective trade patterns, the general cargo fleets of the two superpowers present a striking contrast. The Soviet Union has the third largest fleet in the world, with 1,712 carrier ships, accounting for 8 percent of the world's general cargo ships and 10.3 percent of the tonnage. With its 425 ships, the United States has only 2 percent of the world's fleet and 4.2 percent of the tonnage. It is a medium-ranking fleet, 13th on the world league table. Indeed, the strength of the United States lies only in the category of container ships. With 126 vessels and an aggregate of 2.9 million metric tons, it has by far the largest container fleet in the world.

Despite this, the present capability of the United States to transport its own seaborne trade is severely restricted. This applies both to trade in general and to strategic minerals. In any evaluation of the defense preparedness of the United States, this aspect of vulnerability must be added to those associated with the sources of the materials and the sea lines of communication. It is clear, therefore, that whereas problems related to sources of strategic minerals and sea lines may well be beyond the control of the United States, its lack of carrying capacity can be remedied.

Not only does the U.S. fleet fall short of its strategic size and structure requirements, but it also happens to be one of the oldest fleets in the world. With 2.5 percent of the ships in the world fleet 30 or more years old, the percentages of the U.S. and Soviet fleets of that age are respectively 22.1 and 1. This U.S. percentage is by far the world's highest. The key factor in this analysis is not only that the U.S. fleet is inadequate, but also that it does not bear comparison with Soviet merchant shipping. Adding together the major categories of ship (i.e., tankers, ore and bulk carriers, general cargo ships, and container ships), the total number of ships in the Soviet fleet is 2,327, while the U.S. figure is 993. It would be somewhat crude to compare the Soviet Union's total trade per ship per annum of about 82,000 metric tons with that of the United States, which is approximately ten times greater. The aggregate gross tonnage of the four categories is, however, similar, at about 15.5 million metric tons. By the same procedure, this gives a figure for the Soviet Union of 13,000 metric tons per gross metric ton of shipping, more than four times greater than that for the United States. While such calculations are relatively inconsequential, they do serve to highlight the differences between the carrying capacities of the two fleets. Furthermore, by virtue of the U.S. dependence on a wide range of trade routes, the disparity in ship numbers assumes a greater significance than the parity in gross tonnage between the two fleets.

However, if in an emergency it were possible for the fleets of the respective blocs to be coordinated, the Western fleet would be at a distinct advantage. The fleet available to the bloc as a whole numbers over 27,000 ships, with an aggregate of over 149 million gross metric tons, accounting for over 35 percent of the total world fleet. These figures highlight the decline of the United States, which accounts for only 24 percent of the ships and 13 percent of the tonnage, even though it accounts for 28.5 percent of the NATO countries' trade. By contrast, the Warsaw Pact fleet accounts for 13 percent of the world's shipping total and only 8 percent of the tonnage. These percentages are dominated by the contribution of the Soviet Union, being responsible for 62 percent of the aggregate trade and, furthermore, accounting for 81 percent of the ships and 73 percent of the gross tonnage.

However, under conditions of tension, at least in the initial stages each country would in all probability have to mobilize its own fleet, with the United States thus unable to count on the fleet of its allies. Japan's position is somewhat different, despite its vast fleet. Its essentially defensive military posture might well limit its scope for action.

CONCLUSION

By virtue of its reliance on external sources for strategic minerals, the United States is dependent on a number of extended sea lines of communication. While those crossing the Pacific are important, the most vital at the present time is the Cape Route from South Africa to the ports on the east coast of the United States.

As a result of this global pattern of routes, shipping has to transit several choke points, including the Cape itself, those associated with the Suez Canal sea line, those in the Caribbean, and the Malacca Strait.

The strategic mineral supplies to the United States are carried along the longest routes in the world, through some of the heaviest shipping concentrations. Furthermore, the capacity of the merchant fleet of the United States is both limited and declining.

NOTE

1. The Third United Nations Conference on the Law of the Sea.

5

Global Threats and Possible Responses

Strategic mineral supplies may be at risk, not only as a result of regional and local issues concerning the supplier countries, but also from the potential vulnerability of the transport infrastructure. There are also threats of a more general, global nature related to competition of varying intensities. These may range from problems arising out of normal trading to the possibility of a "resource war." Trading can be disrupted not only through the deliberate acts of governments or multinational corporations, but also as a result of the routine operation of the world economic system. Premeditated actions can be multilateral or can be implemented by groups such as associations or cartels. Certain economic, political, and military activities of the Soviet Union, particularly in the field of strategic minerals, have been characterized by some commentators as a "resource war."

RESOURCE COMPETITION

While competition for resources is by no means confined to the superpowers, that between the East and the West is of greatest significance. It is often suggested that the Warsaw Pact countries as a whole are virtually self-sufficient in strategic minerals, but there is some evidence (especially from trade returns) to suggest that this may not be the case. While import dependence exceeds 50 percent only in the case of fluorspar, the Soviet Union is also a major importer of four strategic minerals. Moreover, it is known to have entered the market at different times with the aim of procuring at least four others. It can be assumed that over time, as Soviet reserves become depleted, the Soviet Union will have to turn to external sources more and more. It is also worth noting that although the Soviet Union has abundant mineral resources, extraction can prove difficult and costly. Therefore, although the strategic needs of the West are undoubtedly more acute, in future there is likely to be greater real competition between the West and the Warsaw Pact countries. The entry of the Soviet Union into the international strategic minerals market can be interpreted as signaling a desire

1. to replace declining home production, or
2. to limit the access of the West to strategic minerals, or
3. to establish links with southern African countries that could, in future, lead to a Soviet monopoly over their strategic mineral exports.

However, mineral experts interpret the behavior of the Soviet Union as a response to temporary imbalances in the domestic demand and supply situation. Certainly, there is no discernible pattern in the actions of the Soviet Union in this regard, with poor planning and infrastructure problems perhaps as underlying causes. Any interference in petroleum supplies would have a more devastating immediate effect on Western economies, although such an overt or precipitate action would have its attendant risks. The Soviet Union has been a supplier of strategic materials to the West for the past two decades and its market behavior has been sensitive not only to price changes, but also to its own hard currency requirements. Its policy of aiding Third World countries supplying strategic minerals has not significantly undermined trade with the United States so far.

Thus, while there is clear evidence of competition between East and West, it is difficult to discern any guiding rationale behind Soviet market behavior.

Some would see a danger of equal gravity in the competition among the Western countries themselves. When severe shortages have occurred in the past (e.g., the 1973–74 petroleum shortage), no united action was taken to deal with the resulting crisis. NATO countries have disagreed among themselves on a number of issues, particularly those relating to energy. With marked discrepancies in their material reserves, the urge to compete often dominates the need to cooperate. This lack of cooperation is intensified by the pursuit of different stockpile policies. It is difficult to believe that, under conditions of scarcity, countries without stockpiles could expect to benefit from the generosity of their more prudent allies.

Where European NATO countries are dependent on the United States for strategic minerals, they would almost certainly experience shortages if domestic demand were to increase in the host country. This has already happened in a number of instances, most notably molybdenum. Until a definite policy on strategic minerals embracing the West as a whole can be forged, such a situation is always likely to recur. Given the dominance of private enterprise in the West, international competition is not the result of governments' policies, but is caused by individual companies, especially multilateral corporations, who fail to respect national boundaries and have a different view of commercial and political interests. As a result, Western governments have attempted, from time to time, to exercise control over the activities of multinational corporations.

Internal competition may also result from competition among investors. The role of the free market in the sphere of strategic minerals needs careful consideration. For example, a large rise in germanium prices resulted not only from the Soviet Union's decision to withdraw supplies and the uncertainty of African sources, but also from the competition among rival investors.[1] Furthermore, such internal competition raises the question of how rapidly the government could, in case of emergency, gain control over privately owned stockpiles of strategic minerals.

Apart from competition between East and West, there is also an element of competition between North and South. Many strategic minerals are produced in developed as well as underdeveloped countries. Not only do producer countries compete in the world market, but they also compete for investment. Although the bulk of present investment takes place in the developed countries, certain underdeveloped countries are critically placed to control supplies of certain strategic minerals to the West. There is no coherent Western policy to take account of this aspect of reality. Moreover, the quest for strategic minerals, unlike that for energy sources, which is likely to ease in the future with greater use of solar power, will continue to be subject to geopolitical constraints.

In the world trading pattern of strategic minerals, different groups compete at different levels during a period of relative peace. However, in times of international tension or conflict, the situation will rapidly deteriorate. Thus, for example, if the United Nations were to mount a successful economic boycott of South Africa, competition would intensify at various levels in the world strategic minerals market.

MULTINATIONAL CORPORATIONS

The adequacy of supplies in relation to demand at the prevailing prices is governed principally by both the level of investment and the behavior of the market. In the short and medium term, costs of mineral extraction are likely to rise. Therefore, in order to maintain existing plants and finance new ventures, a higher level of investment is required. However, two factors tend to militate against a sudden increase in investment. Both the fear of insecurity and an increase in costs at a higher rate than that of prices might have the same prohibitive effect.

Globally significant mineral extraction requires either government or major corporate finance or a combination of the two. This raises the important question of how the operation of multinational corporations affects both world trade and, on the local scale, the point of extraction.

Multinational corporations can be defined as centrally controlled enterprises that control assets in two or more countries (Barnet and Muller 1975). With perspectives transcending national boundaries (Radice 1975), their decisions tend to favor the corporate rather than any particular national interest. Indeed there is considerable potential for conflict in this difference of interests. While national economic concerns might include foreign-exchange requirements and employment levels, in addition to profit, corporate policies tend to be dominated by a desire for global profitability. A number of Third World countries have experienced a clear disparity between their political independence and economic autonomy. Those possessing strategic minerals view them primarily as a development-generating resource rather than as raw materials integral to the world industrial system.

The relationship between the multinational corporation and the host country goes through two main stages. At the beginning, due to the high cost of exploration and establishment of the industry, the multinational corporation is well placed to dictate its own terms. However, the high fixed costs initially fundamental to the corporation's bargaining power eventually become the main source of vulnerability.

This process was particularly pronounced in the period following World War II. Countries in which strategic minerals were abundant developed strong nationalistic tendencies and the attendant demands for rapid modernization were great. At the same time, they became more sophisticated in their dealings with multinational corporations. Their trained work force was deployed to monitor the activities of multinational corporations. They were able to generate sufficient pressure in favour of renegotiation of terms. In a number of cases, effective control passed from the corporation to the government of the host country.

On the other hand, the nationalization of mineral production by such countries as Mexico, Zaire, Zambia, Chile, Peru, and Guyana might also upset market trends. With the rise of inflation and the deepening of the recession, "creeping expropriation," once feared by the multinational corporations, has been less frequent since the mid-1970s. Direct confrontation between the host countries and multinational corporations has, in a number of instances, given way to mutual accommodation. The developing countries need investment. A United Nations commission that enquired into the impact of multinational corporations on international relations has influenced the drawing up of guidelines for negotiations and agreements between the two sides (GJW Government Relations 1981).

Increasingly, host governments are participating in the working of multinational corporations, while state-owned companies control half the Western world's bauxite and copper supplies as well as significant amounts of tin, lead, and zinc production. However, far greater amounts of investment are still being directed into the more secure producer countries such as the United States, Canada, Australia, the Philippines, and even South Africa, rather than to countries such as Chile, Peru, Zaire, and Zambia. Thus multinational corporations not only affect the pattern of world trade, but possess a great potential for influencing its direction and volume.

Although multinational mining corporations are best known for their much larger output of base metals, many include strategic minerals in their portfolio. For example, General Mining Union Corporation: chromium, manganese, platinum-group metals, tin, and titanium; Newmont Mining Corporation: ferrovanadium, manganese, cobalt, and molybdenum; Inco: nickel, cobalt, and platinum-group metals; and Malaysia Mining Corporation: columbium, tantalum, tin, and tungsten (Hargreaves and Fromson 1983).

THE RESOURCE WEAPON

In the extractive industries, operations tend to be large scale in character and lead times for new developments or for reopening of old plants are relatively long. When to these factors is added the concentration of sources, producers are well placed to control world trade, either by restricting production or output (and thereby raising world prices) or by deliberately routing supplies to chosen destinations. Action might be taken by a particular country though initiatives coming from a cartel are likely to be more effective.

While a price rise resulting from the actions of individual countries or cartels has the potential to disrupt world trade, partial or total embargoes have the effect of restraining international movement. Embargoes may be imposed on imports or exports in response to the requirements of domestic or foreign policy. Recently the embargo placed upon foreign countries in order to influence their behavior has become highly important. While Arab oil has already been used in such a manner, perhaps more relevant was the United Nations embargo (1966–79) on Rhodesian chrome. Clearly this is a potent weapon that can be used to influence policy, especially if the user has a near monopoly.

In the case of a number of minerals, imports and exports are controlled almost exclusively by two or three countries, which are thus in a strong position to block exports to a particular consumer.

Again, in some cases, unilateral activity could also be effective, although in general the damage caused by embargoes on most nonfuel minerals would be comparatively small and effective only in the short run. Thus, for example, when the Soviet Union reduced its shipments of manganese and chrome to the United States (from 1948 onward), Indian manganese resources were tapped. However, as with the operation of multinational companies, the use of cartel strength and the imposition of embargoes can alter investment patterns, trade in a particular mineral, and, therefore, the overall geopolitical equilibrium (Eckes 1979).

An embargo may also be backed by military force, or the world trading pattern may be disrupted by military action. However, the main danger to the political stability of the major Third World mineral producers appears to lie in changes of ideological alignment rather than in direct military intervention. While it is true that such countries would need to sell their minerals in order to develop, internal struggles and wars are likely to affect trade and could result in significant damage to existing plant and infrastructure.

Thus even though a particular country can impose an embargo on the sales of a mineral to another, given the range of sources generally available and most producers' inability to exercise control over transportation, its effects are unlikely to continue beyond the short run. Furthermore, the economy of the country imposing the embargo seldom has the ability to withstand the pressures caused by loss of export revenue. The only countries potentially capable of imposing an embargo on the United States with any degree of success are South Africa (for chromium and platinum-group metals), Zaire (for cobalt), and perhaps Brazil (for columbium).

As a consequence of OPEC's policy during 1973–74, a number of economies were seriously affected, while at times even the stability of the world economic order seemed in danger. The infuence of OPEC's policy on economic and geopolitical thinking was so profound that there followed immediate speculation about the possible formation of other cartels, particularly for strategic minerals.

The likelihood of exporting countries establishing successful cartels depends, in theory, on (a) economic and political factors including the proportion of the world exports; output and reserves controlled; supply and demand elasticities; the political, ideological, and economic outlook of exporters; and the level of cohesion between the would-be cartel members (Pirages 1978); and (b) mineral factors such as the relative importance of the mineral, the possibility of substitution, world consumption patterns, consumer vulnerability, and the geographical distribution of production. It is clear that in the case of OPEC, whichever criteria are applied, there was only limited homogeneity (geographical, political, or economic) and therefore a key influence must have been the overall strength of purpose of the Middle Eastern countries, with Israel the probable catalyst. However, potential cartels in strategic minerals would differ substantially from OPEC in a number of significant respects.

While strategic minerals are less easy to substitute (particularly in the short run), it is often possible to delay consumption and, in certain cases, to increase production through recycling. Furthermore, producers are generally widespread, representing a wide range of political stances. For example, the Soviet Union, South Africa, Gabon, Brazil, Australia, China, and India are major producers of manganese. Also, in the case of strategic minerals, there is a certain lack of vertical integration between the mine and the market. For these reasons it is unlikely that the success of OPEC will be repeated by the producers of strategic minerals. Nevertheless, the volume of world trade is considerably less than that of oil with each particular mineral correspondingly of lesser importance. Finally, since the increase in petroleum prices also affected the mineral producers, it would be even more vital for them to sell their output.

Associations of major exporting countries already exist for copper, bauxite, mercury, and tungsten, but so far they have not been successful in capturing monopoly rents. Eleven major producers belong to the International Bauxite Association, accounting for over 80 percent of world trade. However, the effectiveness of any cartel is limited by the marginal strategic importance of aluminum and the large global potential for bauxite production. Furthermore, world production is tightly controlled by six multinational corporations that can exert considerable control over prices. In addition, there is a strategic stockpile of bauxite and alumina in the United States that can be used to regulate trade. These examples illustrate the principal reasons for the low level of cohesion among producers and the inability of newly formed cartels to last more than five years.

However, in those specific cases in which the number of import producers is limited, effective cartelization may well be possible. For example, over 96 percent of chromium reserves are located in the Soviet Union, South Africa, Zimbabwe, and the Philippines, while 98 percent of platinum-group reserves are found in South Africa and the Soviet Union alone. For these minerals as well as for cobalt, the number of major world producers is restricted. Manipulation of trade is therefore clearly possible in the cases of chromium, cobalt, manganese, the platinum-group metals, and tin. For example, the Soviet Union is an important producer of all these minerals except cobalt. An alliance between South Africa, Zimbabwe, and the

Table 5.1
South Africa and the Soviet Union as World Strategic Mineral Producers: Production (and Reserves) (1984)

Strategic mineral	South Africa (%)	Soviet Union (%)	Combined Production (%)
Platinum Group	39 (81)	55 (17)	95 (98)
Manganese	14 (41)	46 (44)	60 (85)
Vanadium	34 (47)	34 (25)	68 (72)
Chromium	21 (68)	34 (1)	55 (69)

Soviet Union could result in an undermining of world trade in chromium. If these three, together with Zaire and Zambia, were to coordinate their efforts against the United States, the latter could be faced with problems relating to supplies of cobalt, manganese, chromium, and platinum-group metals.

Despite the possibilities outlined above, no cartel is likely to have more than a short-term effect, with consumers likely to turn to safer, higher-priced deposits, while simultaneously reorienting research and development toward the production of substitutes. In the long term, such measures would be damaging to the exporting countries. Therefore, while there might be short-term difficulties related to specific regions exporting strategic minerals, major confrontations in strategic minerals trade are unlikely. Unilateral protêst action such as that reflected in Jamaica's policy of increasing taxes on foreign-owned bauxite mines is probably more likely than multilateral action.[2] Even unilateral action is unlikely to generate long-term repercussions, because the country initiating it may price itself out of the world market. A recent example of this can be seen in the cobalt prices demanded by Zaire, which lead to a decrease in demand. In any case, the flow of strategic minerals is likely to be interrupted from time to time by such unilateral action, since they represent important developmental assets to most Third World countries.

Although it is unlikely that strategic mineral cartels will emerge, one particular alignment cannot be totally discounted even though it may not appear imminent. As the two leading producers of four key minerals—platinum-group metals, manganese, vanadium, and chromium—South Africa and the Soviet Union between them dominate the world market. This is readily apparent from Table 5.1. A situation might well arise in the future in which a black South African government with friendly ties with the Soviet Union might agree to a joint policy between the two countries that could drastically reduce the access presently enjoyed by the United States to South Africa's minerals.

"RESOURCE WAR"?

The scenarios outlined earlier point to an assessment of the extent to which the trading posture of the Soviet Union in strategic minerals could lead to a future "resource war" (e.g., Legvold 1983).

A "resource war" can be defined as the acquisition of resources by noneconomic means. It therefore tends to destroy the normal trading pattern and economic infrastructure. A confrontation short of actual war will often ensure a resource war. As an aspect of foreign policy, a resource war is aimed at controlling supply sources by using political rather than economic or even military techniques. It is often covert, low cost, and low profile and cannot therefore by clearly identified. As a result, there can be no reasoned response to it. However, both superpowers can and do use such means, depending on the circumstances. The capability of the Soviet Union to undermine the strategic mineral supplies to the United States by such means ought now to be examined.

In order to examine the current situation with regard to the Soviet Union's posture, it is important to separate actual events from potential occurrences. The potential may be considerable, but tangible conclusions can only be drawn from concrete evidence. If a resource war is to be pursued on the basis of concerted acts (or actions), irrespective of whether they form part of a coherent policy, these acts must be analyzed. A particular strategy may be planned or haphazard and opportunistic. According to some commentators the Soviet Union appears to thrive on "managed instability" (Anderson and Blake 1984). On the other hand, planning may be so broadly based that it can take advantage of all available opportunities.

Where a coherent policy seems to underlie Soviet economic, political, and military actions, its overall objectives ought to be considered. Minerals experts associated with the London Metals Exchange do not perceive a general Soviet strategy and characterize most of the actions undertaken by the Soviet Union as being calculated to ensure domestic consumption. Diminishing resources within the Soviet Union, conservation policy, problems related to extraction, errors in the execution of policy, and bad planning can lead to temporary shortages.

There are those, however, who attribute to the Soviet Union a deliberate policy, designed primarily to secure political gains and to cause difficulties that lead eventually to Western deprivation of strategic mineral resources. These two much discussed approaches have come to be termed "access" and "denial." Even if no clear policy can be discerned and Soviet behavior is thought to be opportunistic, the objectives, whether access or denial, would still need to be investigated.

Furthermore, access or denial can be viewed as short term or long term, temporary or permanent in character, perhaps initially resulting in short-term economic market gains, but leading eventually to permanent access. Similarly, Soviet policy may result in temporary political gain from the economic problems of a Third World state, which may in turn lead to a reorientation of trade away from the West.

It might be added that these ideas are not mutually exclusive and may indeed be difficult to separate. The Soviet Union's access may result anyway in Western denial. With a policy of denial, the Soviet Union may have to import over and above its own mineral requirements and may stand to make some commercial gains.

Since, therefore, it is largely a question of Soviet motives, exact definitions are particularly elusive. However, broadly speaking, "denial" is viewed as deviant behavior whereas "access" is not. At the same time, in unraveling the different strands, the analyst may find that small events take on a cumulative significance. Arguments in favor of either policy necessarily have to take account of a mixture of unknown quantities—principally Soviet motives—and known elements such as documented events though these may be subject to a variety of potentially conflicting interpretations.

The argument for denial revolves essentially around the potential for denial. The Soviet Union has been active in southern Africa, the principal source of strategic minerals. However, while this has not taken a direct military form, the Soviet Union does support some of the national liberation movements active in the region. It has been posited by some that the infrastructure (in particular, the Cape Route) involved in bringing strategic minerals to the West would present the Soviet Union with opportunities for denial. It is equally clear that the invasion of key mineral sources or the interception of supplies en route would be tantamount to an act of war and could therefore lead to escalation. Thus the Soviet Union would benefit little beyond the period of tension preceding the eruption of war. This does not rule out the possibility of Soviet engagement in a variety of activities of less extreme immediate consequence, such as the creation of an environment unfavorable to the normal conduct of trade.

A coherent Soviet strategy of denial therefore would be difficult to discern, whereas events, usually locally inspired, do occur for a variety of reasons, often connected with racial, ethnic, secessionist, boundary, and industrial tensions.

If access is the crucial factor, it would constitute an indication of the gradual transformation of the Soviet Union from a state of self-sufficiency in strategic minerals to one of dependence. As problems of extraction intensify, costs escalate, while there is the added strain of supplying East European countries with most of their strategic minerals requirements. A slight shift in the Soviet trade pattern in strategic minerals can be discerned, although it is premature to distinguish definite trends. Recent export figures for platinum, titanium, and gold appear to be lower, while imports of cobalt, tungsten, tin, and barite have risen. At the same time, the added requirements for cobalt, tin, and tungsten could lead to more acute competition between the East and the West.

Owing to its limited access to hard currency, the Soviet Union is not well placed to take a full part in normal world trade. It has shown a tendency to enter into barter agreements and, where appropriate, to use its economic and political influence to satisfy its import requirements. In Western countries this aspect of Soviet trading behavior has been cited as evidence in favor of increasing the Soviet Union's access to Western technology in order to encourage it to follow orthodox trading practices.

In the future, access will be a Soviet requirement, but at present there is little evidence to suggest that it is a definite policy. The Soviet Union's market activity is not guided by a clear strategy and is probably the result of temporary imbalances or poor planning.

Concrete evidence suggests that the current strategy of the Soviet Union is marked by opportunism, aimed at maximizing short-term economic gains and ensuring longer-term political benefits. However, the major examples of such activity have had no lasting effects, though their potential significance should not be underestimated. Thus, following the Western embargo on Rhodesian chrome ore the Soviet Union was able to increase profits by buying and remarketing the ore. It has also been noted that Soviet market activity prior to 1979 resulted in a cobalt stockpile able to withstand the effects of the Shaba (Zaire) insurrection in that year.

If the current trends in its trading pattern continue, the Soviet Union may require access. It is, however, clear that there is no dearth of opportunities for political maneuvering in the future. A supercartel, including the Soviet Union and a future (or even the present) government in South Africa would radically affect supplies of chromium and platinum-group metals to the United States as well as influence manganese trade. It must also be remembered that the United States is directly (though only partially) dependent on the Soviet Union for chromium and its dependence on the same source for palladium is annually increasing.

Finally, however, it must also be noted that shortages can occur as a result of the normal working of the economic system at the global level rather than through any deliberate acts by interested states. Cyclical volatility is caused primarily by the mechanics of the economic system. Short-term imbalances between supply and demand create fluctuations in output, profits, and prices, but these are likely to have only a temporary effect. The chief causes are the low, short-run price elasticities of both supply and demand for strategic minerals, all highly volatile and many important for the automobile, machinery, and electrical industries. This invariably results in a change in overall economic activity in the course of the business cycle.

POTENTIAL RESPONSES

In the short term it may be possible to overcome shortages in a number of ways, the most obvious being through international agreements or the use of stockpiles, strategic or economic. In extreme cases, countercartels or even military action might be contemplated. Technological advances in the fields of substitution, recycling, and conservation probably hold the best prospects for the short or medium term. In the long run the situation could be improved by subsidizing domestic production or, where possible, increasing the range of foreign sources. Indeed, there must be some scope for foreign policy initiatives, including the provision of aid, to influence favorably the stability and reliability of current sources. In order to ensure a secure supply of strategic minerals in the long term, exploration of new sources, domestic and foreign, should be considered, together with offshore mining.

The study of economics is not yet sufficiently advanced to allow the development of macroeconomic stabilization policies capable of controlling the business cycle (or cyclical volatility).

Nor is the world at a stage at which associations between major importing and exporting countries can be easily established as a prerequisite for such developments. International commodity agreements through such associations could create conditions in the short run under which buffer stocks and export controls could be used to reduce the adverse effect of economic factors. Such associations would act as forums for discussion and discourage unilateral action or activity or domination by cartels. The major objectives of producers and consumers—equitable prices and stable supplies—may thus be achieved. Important among such forums already in existence, albeit not yet fully established, are the Commonwealth Prime Ministers' Conference, United Nations Conference on Trade and Development (UNCTAD), the Paris Conference on International Economic Cooperation (the North–South dialogue), the European Community–African, Caribbean and Pacific countries (EEC-ACP) negotiations, and the United Nations Conference on the Law of the Sea (GJW Government Relations 1981). These bodies have contributed to a growing awareness of the interdependence of the North and South and also of the need for stabilization of exports. Fluctuating production and therefore fluctuating prices not only have an effect on the consumer, making planning difficult, but also retard the development of the producers.

Multinational corporations are better placed to adapt to such changes than the state-run companies (such as those in Zaire, Zambia, and Chile, for example). With greater stabilization, the geographical horizons for the exploitation of strategic minerals would be widened. Whether an agreement similar to the Stabilization of Exports (STABEX), signed in 1979, could be established for strategic minerals is still a matter for conjecture. It is, however, important to note that the rate of growth of less-developed countries is greater than that of developed nations. Indeed, by 1990 the former will be responsible for approximately 25 percent of the world economic activity. Therefore, partnerships between customers and producers, governments and multinational companies are increasingly likely to be formed.

STOCKPILES

The potential benefits of stockpiling as a means of stabilization are worthy of serious consideration, even though various problems arise from such a policy. At the outset, it is important to distinguish between economic stockpiles and strategic stockpiles. Strategic stockpiles are retained specifically to combat political and, in particular, military intervention. Economic stockpiles, however, should be carefully monitored to prevent market distortion. The stockpile should follow price changes and not fix the market trend.

Part of the current stockpile debate occupying planners, particularly in the United States, concerns the practical problem of distinguishing between the possible uses of the two types of stockpile. For example, in peacetime it would seem reasonable to treat an embargo on strategic mineral supplies as a hostile act and to use the strategic stockpile, even though its stated function is to allow orderly mobilization for war. Obviously, the longer a conflict continues, the more important become the industrial bases of the protagonists. Therefore, essential civilian as well as military stockpile requirements would need to be taken into account, together with possible developments of substitutes and upgrading of recycling facilities.

Even if there is no intention to use the stockpile economically rather than strategically, problems can still arise from changing requirements. The inventory of the stockpile must reflect the changing needs of industry and this must be carefully managed in order to prevent surpluses undermining the market. Not only should the types of material held be carefully considered, but also the form in which they should be stockpiled.

Just as processing capacity has declined in a number of NATO countries, so unprocessed minerals have become obsolete. The lack of such industrial capacity intensifies the vulnerability of these countries, yet they do possess some advantages. As the suppliers are drawn more into the processing industry, their returns become greater and an interruption of supply would be much to their disadvantage.

Moreover, since the processed mineral is more standard than the raw material, downstream industrial facilities no longer need to be geared to one specific source of supply. Thus stockpile management, including defining the materials required, establishing goals, checking inventories, and monitoring deterioration, is a sophisticated procedure. Furthermore, unlike most other areas of government funding, a stockpile is unlikely to be a depreciatory asset. Through time, mineral prices will probably rise. Indeed, in the event of their being required in an emergency, they would almost certainly fetch a price substantially greater than that at which they were purchased.

Private stockpiling by user companies is also an option. With the support of the government, this might be raised to a required level (Holt and Stanley 1983). It has been suggested that the existence of a government stockpile tends to decrease motivation for private stockpiling. However, given the potential complexities of allocation from such a stockpile, this opinion appears somewhat naïve.

The merging of the strategic stockpile with private stockpiles would present considerable administrative problems. To maintain checks on inventory levels, to control the system in a manageable form, and to identify abuses would be an extremely complex exercise.

TECHNOLOGICAL MEASURES

Research and development can, in the short to medium term, reduce the vulnerability of a country to shortages. By improving each stage of the extraction and fabrication process, conservation would be significantly improved. Yet the most important savings would probably arise from research into substitution and, to a lesser extent, recycling. Since the immediate benefits are likely to be limited, the government would need to provide some support for such work. The outlay would, however, be unlikely to exceed the requirements for maintaining the stockpile or for subsidizing domestic production. If necessary, improved technology could be stored in a technology stockpile, ready for use in case of emergency. The existence of such a stockpile would also help guarantee confidentiality and encourage corporations to donate the results of their private research.

Substitution covers a wide field and can be temporary or permanent, reversible or irreversible. It can refer to any stage in the manufacturing cycle from total replacement of the raw material to complete change in end use. As a consequence, total or partial substitution may be either immediately cost effective or form part of contingency planning. Perfect substitution of one material for another is by definition impossible. Furthermore, any two substitutes are bound to differ in cost or efficiency. While in some cases the impact of substitution on long-term performance or even resistance to the environment may be minimal, difficulties regarding simple engineering may well arise. At the same time, it is possible that substitution may affect not only design, but also strategic vulnerability. In certain cases, nickel might be replaced by molybdenum, though pressure on supplies of the latter will inevitably increase as a consequence. Recent research has concentrated on ceramic materials and strategic metals employed in the form of spray or molecular thickness, with the aim of decreasing overall requirements.

Many of the problems outlined can be illustrated with reference to the aerospace industry, in which each engine alloy represents a sophisticated and complex balance militating against substitution except over a fairly long time span. In the case of key engine alloys, which would require rigorous retesting, this is at least five years. Furthermore, reliability is a sine qua non for any change, either in the aerospace industry or in the manufacture of military equipment. The fact that the criticality of a particular alloy is closely related to its use would create difficulties wherever the question of substitution is considered. As higher engine thrusts are achieved, the environment in which the equipment has to operate becomes increasingly hostile and substitution must satisfy more and more stringent criteria.

At the present time, engines require a running time of 150 hours for type testing before they can be sealed and certified. Any changes would require a further type test to ensure that the new alloy is a satisfactory substitute. At a minimum, type testing could take several months and consistent changes in overall consumption pattern may not be clear in less than four or five years. Complete changes could therefore take up to ten years, although in emergencies (e.g., war) standards would have to be altered and time intervals reduced.

Substitution may also be constrained by: (a) difficulties such as inertia and investment provision within the industrial structure itself; (b) social factors ranging from employment provision to pollution; and (c) psychological problems related to resistance to change. Recycling can also affect supply, although with the introduction of new technology (e.g., pressure molding and the use of power alloys) the amount of new scrap has been significantly decreased. With regard to old scrap (in particular, scrap derived from alloys) containing only a minute proportion of the particular metal intended for recovery, the recovery industry is developing a whole new range of techniques. Examples include the recovery of germanium from flue dust and other metals such as vanadium from pulverized ash. Since the transport costs are minimal and the material is already in a form suitable for mineral extraction, far lower concentrations than those normally required in the ore can be considered potentially viable.

NEW SOURCES OF RAW MATERIALS

Even among the major consuming countries such as the United States, the full potential for resources has not been assessed. This may be accounted for by a number of factors, including difficult terrain, hostile climate, lack of infrastructure, and the influence exercised by environmental interests. New sources in the producer countries will also become available and it must be remembered that extraction from the land will continue to be cheaper than underocean exploitation. In such countries as Brazil, the Sudan, and China, geological surveys are far from complete, while new discoveries have been made even in developed mining areas such as Zimbabwe and certain Andean states (e.g., new low-grade copper ore bodies are discovered annually in the Andean region) (Anderson 1982).

The development of offshore resources is of long-term significance, and the United States expects to commence exploitation in the early 1990s. With the advanced technology required to recover the seabed nodules and to separate the constituent minerals, offshore exploitation will be of great benefit to the West in general and to the United States in particular. Manganese will no longer be considered vulnerable, and supplies of copper, nickel, and zinc will be strengthened. Nodules recovered from the Indian Ocean are known to have a high concentration of cobalt. Attention has recently been focused on polymetallic sulphides containing a greater variety of minerals, the exploitation of which may not present difficult technological problems. However, it is unlikely that deep-sea mining will be begun in a systematic manner during the current century, not only because of the costs involved, but also because of its effect on the economies of many Third World mineral producers. The exploitation of brines (notably in the Red Sea) and marine muds is at present in its infancy, although it holds promise for the future. It is already known that economic quantities of zinc and silver can be recovered from Red Sea brines. Shallow-water technology is already available, but as in most cases involving offshore extraction, scientific research is more advanced than agreement between nation states on the political and legal aspects of international boundary demarcation.

POSSIBLE RESPONSES BY THE UNITED STATES

In view of the vulnerability of the strategic mineral sources of the United States, various measures have been considered to overcome potential crises of availability of important requirements. Some of these have become part of government policy and are adumbrated here.

NATIONAL DEFENSE STOCKPILE

In the United States, the NDS constitutes the major means of reducing dependence on foreign strategic mineral supplies. It was set up by the executive under the provision of the

Strategic Minerals Act (1939). Its existence was formally legitimized by Congress under the provision of the Strategic and Critical Materials Stock Piling Act of 1946, Public Law 520 (American Mining Congress 1963). The acquisition and supplementing of inventories acquired before and during World War II was authorized; the act specified that the NDS should be held only for meeting national emergencies. To the provision for building a military stockpile under the act have been added:

1. The Defense Production Act inventory (authorized by Congress in Public Law 774, 1950) to help meet the requirements of the Korean War emergency
2. the supplemental stockpile (authorized by Congress, Public Law 480, 1954 and Public Law 540, 1956) to enable bartering of perishable agricultural commodity surpluses in exchange for foreign strategic minerals.

Some of the commodities obtained under both forms of supplement have subsequently been transferred to the defense stockpile.

The NDS is the cornerstone of U.S. minerals policy (Congressional Budget Office 1983). While the stockpile itself is managed by the General Services Administration, the Federal Emergency Agency (FEMA) is responsible for policy and planning related to it. FEMA establishes the goals through a process principally involving the following departments: Defense, State, Commerce, Interior, and Treasury. Discussions taking place in FEMA include the simulation of wartime minerals requirements as well as domestic primary and secondary production under conditions of normal mobilization, together with a number of emergency scenarios. A quadrennial total review is undertaken by the National Security Council (NSC). The 1985 appraisal has been completed and is still the subject of Congressional attention.

The NDS was built after World War II with a view to dealing with a five-year war scenario. It was worth approximately $9 billion. In 1958 the time span of the projected war scenario was reduced to three years. This continued until 1972. The value of the goals fluctuated between $11 and $13 billion. During the 1973–75 period, the time span was further reduced to one year. During the period 1958–75 as a whole, as the downward adjustment of the time span progressed, the inventory value of the NDS represented a sum of approximately $7–$10 billion in excess of the corresponding projected goal value. From 1975 onward, contingency planning has once again been extended to cover a three-year war scenario, involving goals valued at sums ranging from $18–$20 billion dollars.

Discussions of goal planning with respect to strategic minerals must include reference to:

1. the origin of supplies
2. the volume of current usage
3. the postulated volume of usage during conditions of surge
4. storage and other management problems
5. secondary production capacity
6. the scale of requirement

Thus at the present time supplies from southern Africa appear to be most at risk, while those from Canada and Mexico seem to be the most secure. A generous insurance against supply disruptions, related to current and possible future needs, is necessary. If, however, the total requirement is small, it should be possible except under the most dire circumstances to obtain supplies.

These considerations highlight several underlying problems of the defense industry of the United States, which seems to have been thrown out of balance and needs to be restored in order to face the problems of surge and lead times in a realistic manner. One of the untoward consequences of this imbalance has been a decline in the capacity of the strategic minerals industry to carry out secondary processing from the level of refining upward. This in turn has led to a shortage of skills in a number of areas and this requires urgent investigation. At the same time, the instability of the market and the lack of incentives have inhibited investment. At the very least, assured supplies must be found for defense and essential civilian use and there must be scope in the future for supplies to be expanded to meet general civilian requirements.

In 1985 the stockpile consisted of 89 commodities, including 75 of mineral origin and 14 intended for agricultural and pharmaceutical use. As one mineral may be stockpiled in various forms, the 75 of mineral origin comprise 34 different minerals (if the platinum group is taken as a single commodity). For example, chromium is entered under the following seven categories: chromite-chemical, chromite-MEL, chromite-refractory, chromium-ferro high carbon, chromium-ferro low carbon, chromium-ferro silicon, and chromium metal. Stockpile goals have been set for 63 minerals, and 13 commodities have been marked for agricultural and pharmaceutical use. No specific goals are defined for sapphire and ruby, silver, and a particular category of morphine. Furthermore, within the stockpile are holdings of 39 nonstockpile-grade commodities, for which there are of course no goals. The inventory of 47 commodities is below the goal level, in ten instances only marginally. For example, in the case of beryllium, holdings amount to 99.9 percent of stockpile goals, while for the three categories of tungsten for which goals are set, the figures vary from 96 to 98 percent. On the other hand, there are no holdings of vanadium-ferro, pyrethrum, natural insulation fibers, or for the two categories of cordage fiber. Fifteen items are shown in the inventory up to 20 percent or less of stockpile goals.

There is, however, an excess over stockpile goals of 21 commodities. During recent years various attempts have been made to sell these, perhaps with a view to using the return to help make good the deficits. To have the total inventory at goal level would require some $20 billion against its current value of $13 billion, a figure that includes excess commodities to a value of $5 billion. Recent purchases have included refractory bauxite, cobalt, nickel, palladium, tantalum, and titanium. Further commodities on order include beryllium metal, iridium, quinidine, and vanadium pentoxide; of these, the last mentioned is of greatest significance.

A further 11 items (four of which there are no holdings and five of which are classified as "other inventories") are listed as the commodities without goals. Of the group of five minerals, for example, lithium is held for the Department of Energy.

Other possibilities discussed at FEMA to ease the supply problems of the United States include:

1. the expansion of the titanium sponge industry
2. the creation of a domestic natural rubber industry using the indigenous Guayali plant
3. the reestablishment of some cobalt mining
4. the development of a home refractory bauxite industry using aluminum oxide from local clays

Under The Omnibus Budget Reconciliation Act 1981, authorization has been given to dispose of silver, diamonds, mercury, antimony, mica, and iodine. However, this inevitably poses problems for the markets in each of these commodities. The arrival of some $800 million worth of silver would of course have an instant depressive effect on the market, and sales of such commodities would have to be conducted on the basis of agreement with the open market. The use of the proceeds from such sales for making good deficits in the strategic mineral stockpile has already received attention. The idea of an independent corporation being appointed to manage the stockpile is gaining ground.

The deficits, excesses, and management of the stockpile raise problems of a broad nature. Despite certain changes, the basic inventory of strategic minerals has remained practically unaltered since World War II, while the advance of technology has considerably altered the requirements of the defense industry. It appears that, in view of the changing specifications of the minerals required, a general tendency to use synthetic commodities more and more, and alterations in geographical vulnerability (and other related variables), a penetrating reappraisal of the stockpile requirements is inescapable. At present, apart from the NSC's appraisal, several studies are being completed on individual items, and these should identify possible additions and deletions. However, far greater activity than is presently the case should take place at the interface between the stockpile and the defense industry. For example, the range of exotic metals now required in high technology and for which the United States is dependent on foreign sources, is hardly recognized as such in the inventory. Furthermore, any inventory should bestow attention on vital fabricated and semifabricated items. At present, the only such commodity listed is jewel bearings; but already research on the defense base could supply a further range of such components that would cause blockages in the system during mobilization.

Apart from these more technical considerations, the distinction is also frequently made between a strategic and an economic stockpile. Clarification is urgently required as to when the stockpile could actually be used. Under present conditions, political difficulties, civil unrest, and general instability, rather than a major war, loom on the horizon. It is by no means certain that interruptions to supplies from such developments would create conditions allowing the legal withdrawal of stockpile materials. However, in practice, the only withdrawals so far made in peacetime have been of mercury during the 1950s and long-fiber asbestos in 1979. In both cases, market considerations lay behind the decision to withdraw.

Finally, as a result of the general economic constraints, other stockpile formats must also be considered. Since the presence of a defense stockpile is thought to militate against private stockpiling, perhaps privatization itself with tax incentives should be considered as a practical alternative. If such a prospect were feasible, problems such as management updating would disappear, since inventories would reflect current market forces. However, to place the safety of the United States during a national emergency in the hands of commercial stockpile holders would be a matter of grave concern in the judgment of many.

RESEARCH AND DEVELOPMENT

Research and development constitute a second and probably less costly approach to the task of lessening the mineral vulnerability of the United States. Possibilities for substitution, recycling, and conservation can be examined at all stages of the process, from the initial mining to the end use. Of these, substitution has received the greatest attention and is regarded as a panacea. In the United States the federal government accounts for about 50 percent of the total expenditures on minerals research and development (Congressional Budget Office 1983). However, such aspects as utilization, material evaluation, and the development of special materials have occupied disproportionate attention. Some scope may indeed exist for diverting such resources to basic mineral development and processing. However, 80 percent of minerals research and development is funded by private firms, and therefore the fields of activity are subject to commercial constraints. Furthermore, considering the outlay involved, basic research findings may be considered confidential. In order to maximize their usefulness, such research results should be accessible in time of emergency; the gathering of information within the School of Engineering at Vanderbilt University is a significant step in this direction.

Recycling processes have been developed to such an extent that no great potential for further advance exists. Metals such as aluminum and platinum are already characterized by a high rate of recycling. In cases such as that of the exotic alloy metals, where recycling is particularly difficult, the costs are likely to be prohibitive. Conservation may offer better prospects for the future. Research and development, particularly in the sphere of substitution, give rise to questions of time, of government support, and of transfer from laboratory scale to production-line scale (Landsberg 1982).

One option to reduce America's dependence on foreign sources that is frequently canvassed but of uncertain value is to increase support for the domestic mining industry through protection, subsidization, or the opening of federal lands. Protection might be achieved through a quota system, breakpoint tariffs, import duties, or *ad valorem* tariffs. However, all these measures would impede the free flow of trade that has been, for a long time, a cornerstone of American diplomacy.

Under The Defense Production Act (1950), a sum of $2 billion was made available to meet the shortages resulting from the Korean War. It has been assessed that, over the following ten years, it led to additional production worth $8.5 billion, most notably of aluminum, tungsten, copper, nickel, and titanium. Furthermore, columbium-titanium mining and processing were expanded abroad. In general, the results were considered to be far too expensive because, in some cases, domestic producers were being paid at almost twice the world market price. When the subsidies ended, production tended to fall rapidly. Title III of the bill, with the exception of one loan (1967), has not been used, although the original act was amended in 1982 with a view to assisting industrial modernization with special reference to defense materials.

While subsidies have proved less than successful in American experience, an initial outlay, particularly given the depressed state of the mining industry at present, could in some cases trigger continuing production. Also, any such scenario must take social and economic benefits into account. When the complete equation, including the vulnerability of the United States, import costs, social benefits, and the rejuvenation of mining communities is considered, the case for supporting the production of certain minerals may appear more justifiable.

PUBLIC LANDS

The third element, that is the opening of public lands, has received much publicity, particularly within the mining lobby. A key issue here concerns the actual proportion of federal land that is closed, partially or wholly, to mining. Indeed, in the president's National Materials and Minerals Program Plan and Report to Congress (1982), an estimate of 40 to 68 percent is given.

The Office of Technology Assessment (1979) gave the following figures for 1975: 127 million acres closed to access for hard-rock mining; 48 million acres high restricted; and 380 million acres moderate or slight restrictions on access. This did not include Alaska in which an aggregate of at least 96 million acres has been withdrawn from mineral use. The National Strategic Materials and Minerals Program Advisory Committee in the Department of the Interior (1985) has recently undertaken an inventory which will be useful for compiling reliable statistics on land closed to mineral exploration.

Not only is access limited to certain areas, but also the minerals themselves are distributed in ways that might further limit it. The scientists' knowledge of deposits is largely speculative, and access constraints prevent detailed exploration. However, using information already available, it should be possible to limit environmental damage during preliminary exploratory and survey work. In this context, it should be noted that present mining disruption affects a total of 0.3 percent of the land area. It must be remembered that rich deposits of the key strategic minerals may not be found, and, even if they were to be discovered, the lead times to their exploitation are likely to be at least ten years.

DIVERSIFICATION OF SOURCES

The diversification of mineral sources is a serious option. A policy of development would militate against embargo or cartel action. In the event of hostilities, diversification would obviously increase the chances of obtaining uninterrupted supplies. However, unless the new sources were located in North America, import reliance and vulnerability would still remain.

It is difficult to find ways of encouraging consumers (e.g., private corporations) to diversify away from the most commercially advantageous sources. An attempt was made with the Western Hemisphere Trading Act (1933), which is no longer in existence. Through American aid it might be possible to upgrade the infrastructure and facilities of other sources, which potential users would then find attractive. However, the fact remains that over the past five years the strategic minerals trade, both in America and in European NATO, has concentrated on a few specific sources. Japan is the only major user of strategic minerals with a successful record of diversification (Anderson 1984).

Any consideration of diversification must also take into account offshore mining. The United States, with its long coastlines facing the world's two major oceans, is particularly well placed in this respect. Though the technology is available, political and legal problems involving its use have not yet been solved. The United States has refused to sign the Law of the Sea Convention (1982), mainly because it rejects the proposed role of the International Seabed Authority. This authority would be empowered to expropriate areas of seabed, technology, and minerals produced. The United States, which is at present the only country capable of exploiting the deep sea, would, like any other member state, have only one vote on this body. Although small-scale exploitations may begin on an experimental basis within the next few years, it is unlikely that commercial mining of the deep seabed could occur before the start of the next century. Until then, land-based resources must remain significantly cheaper than anything that could be produced from such a difficult environment.

Many of the foregoing arguments of course affect and are affected by foreign policy. A sharper focus on the foreign policy aspect of the strategic mineral policy of the United States would be appropriate. The underlying cords of dependence binding the United States to a vast range of foreign sources might be strengthened by a more overt diplomatic and foreign aid program. This could be directed through the World Bank, but government-to-government aid must also be seriously considered. South Africa already occupies a permanent place on the foreign policy agenda of the United States, but the stabilization and diversification of sources elsewhere would almost certainly require the leadership of the United States and could offer the least expensive long-term approach to alleviating the problems of strategic vulnerability faced by the United States.

NOTES

1. This information was obtained from a broker on the London Metals Exchange and probably refers to the period 1981–82.
2. In 1974 the taxes increased by 700 percent.

6

The Strategic Minerals Problem: Summary

This study has illustrated in some detail the current dependence of the United States on external sources of strategic minerals, but the actual picture is an ever-changing one. For example, within a short time, tin has changed from a strategic commodity with definite potential for East–West geopolitical competition to a metal of decreasing criticality of which there is, at the moment, an oversupply. Since every problem identified is governed by a unique combination of many variables, even seemingly slight changes can produce dramatic results. Therefore, the geopolitical scene with regard to defense-related mineral supplies must be characterized as unstable.

Problems related to data supply exacerbate the overall situation, because a general lack of defense base data precludes all but a tentative definition of criticality. As has been indicated, this position is now changing and indeed preliminary studies of material requirements have already begun for the Strategic Defense Initiative (SDI). Since so many defense components are intimately related to the rapidly advancing frontier of high technology, the task of precisely identifying the requirements is likely to become gigantic. Even where data are known, the charting of data flows through the political system is likely to reveal many blockages and time lags. Furthermore, the system itself is so complex that deliberate and accidental distortions of data are likely to occur. As a result, perceptions of decisionmakers may become increasingly distanced from real-world facts.

The crux of the argument centering on resource vulnerability relates, principally, to levels of potential. There is only a very limited history of disruptions from which lessons can be drawn. As the gap between domestic resources and defense requirements widens and as world relationships in the economic, social, and geopolitical spheres deteriorate, the probability of disruption will increase. The retention of a balance between the many variables becomes more and more precarious under such conditions. At present only minor irritants are at play, but even a slight variation in the contributory factors could at any moment bring major problems into view.

These conclusions must be viewed against a backcloth of certain philosophical questions, of which an important one concerns free trade. Whether it is indeed "free" and whether it can continue in such a form in a more hostile geopolitical environment remains to be seen. This study also throws into sharp relief a dichotomy characterizing many of the arguments, which can be crudely identified as "private versus public." There can be weaknesses in an insistence upon private remedies for public problems. This dichotomy is pervasive in a number of the problems affecting the supply of strategic minerals.

Several more narrowly based specific conclusions can also be drawn from the study. Using the criteria of criticality—marked import dependence and concentration on a few sources of supply—tantalum, columbium, manganese, cobalt, and chromium are currently of strategic importance for the United States (Figure 6.1). Just below this threshold of what constitutes strategic minerals, but increasing in importance, are tungsten (the use of which is growing rapidly) and the platinum-group metals (which are assuming more and more critical importance). Vanadium, titanium, tin, bauxite, and fluorspar are marginally critical, but, for various reasons, not yet considered vulnerable. At present, tin fails to satisfy many of these criteria, largely as a result of the favorable supply position, but because it presents particular geopolitical risks it is retained in the list.

Figure 6.1
Pattern of U.S. Vulnerability: Strategic Minerals

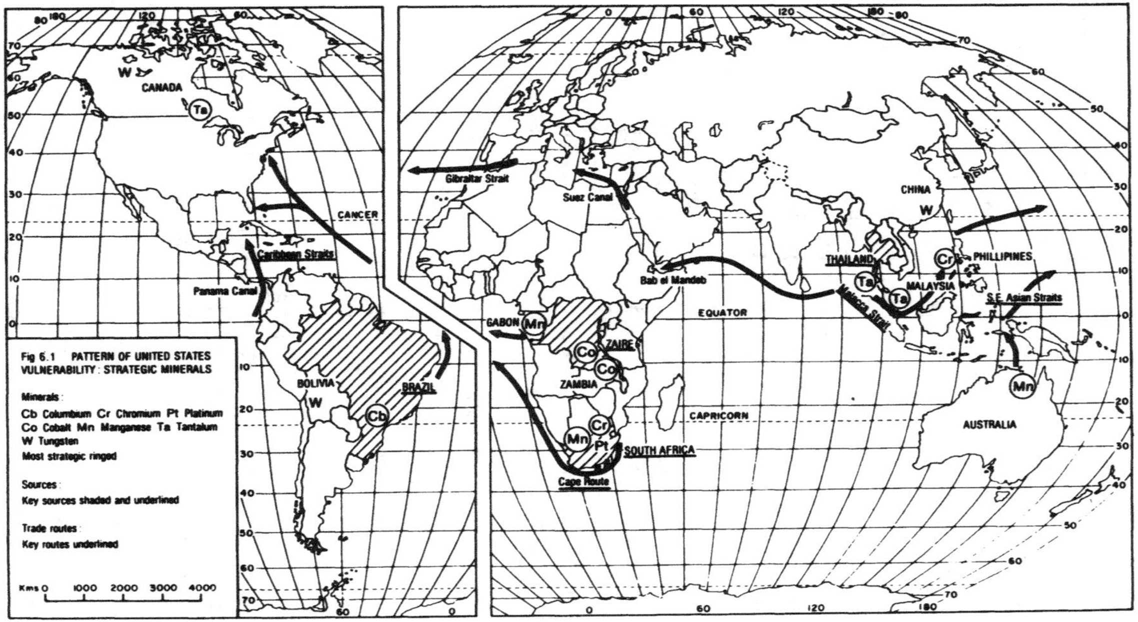

The only major supply source at risk, but in many ways the most important, is South Africa. As internal political changes gather pace, at least a short-term cutoff of supplies might be envisaged. Of the eight key minor suppliers, Zaire and Thailand are undoubtedly most at risk, while the columbium mines of Brazil represent the most extreme concentration of sources. However, given the range of influential factors on a number of different scales, the United States could experience a rapid decline in the reliability of strategic mineral supplies from any of these eight countries. On the other hand, in a reasonably brief period (say, five years), the critical requirements of the United States might change in such a manner as to alter the list of vulnerable countries.

It has been shown, by applying the risk analysis procedure, that threats to supplies vary from country to country, from region to region within a country, and from place to place within the same region. Since each concrete situation has its own unique circumstances, it is extremely difficult to generalize. However, the study indicates that disruption is more likely to result from local events rather than from international events. Thus damage to the mines and infrastructure, resulting from internal conflict, seem to pose the major threat to mineral supplies to the United States, especially those from South Africa. A second and much less probable risk is likely to arise from competition for resources between the two blocs and the operation of the global commercial infrastructure.

Threats associated with sea lines of communication are of a far lower order than those related to sources of supply. Except during a period of increasing global tension between the superpowers, any disruptive action is likely to be locally inspired. The most effective location for such activity could well be choke points, but, in the worst case, disruptions to supplies would only be temporary. During periods of tension, the degree of vulnerability would depend on the precise nature of the war scenario.

Evidence for a "resource war" is limited, but vigilance is obviously required. Precisely because there are divergent views on the subject within the United States, care must be taken to retain all the evidence under review, rather than to jump to conclusions at this stage.

Of the available responses to challenges in this sphere, the most obvious and effective in the short term is provided by stockpiles, strategic or commercial. Increasingly, research and development will provide a wider range of options, but such scientific endeavor must be viewed against a constantly changing pattern of technical and other requirements. Such scientific progress is obviously related to commercial need and, until the potential for disruption of supplies looms large, it will tend to remain dormant. Apart from concrete measures, a penetrating view of both domestic and foreign sources is required for an understanding of the true complexity of problems relating to the acquisitions of strategic minerals.

References

Adams, R. L., B. A. White, and J. S. Grichar. 1979. *Developing a Critical Materials Index: a Pilot Study*. Washington, DC: Department of the Interior, Office of Minerals Policy and Research Analysis, 52.

Advisory Group for Aerospace Research and Development. 1984. *Materials Substitution and Recycling*. Paris: North Atlantic Treaty Organization, 306.

Alonzo, M. V. 1980. The Economics and Finances of U.S. Mining: Crisis and Remedy. *Materials and Society* 4: 25–44.

American Mining Congress. 1963. *Stockpile Study*. Washington, DC: American Mining Congress, 87–91.

American Mining Congress. 1982. *Memorandum* Re: Reagan Administration National Materials and Minerals Program Plan and Report to Congress. Washington, DC: American Mining Congress, 8 April.

Anderson, E. W. 1982. The Need to Safeguard the Supply of Strategic Minerals to the West. In *Global Collective Security in the 1980s*, edited by G. Stewart-Smith, 67–79. London: Foreign Affairs Publishing Company.

_______. 1984. Factors Affecting the Supply of Strategic Raw Materials with Particular Reference to the Aerospace Manufacturing Industry. In Advisory Group for Aerospace Research and Development, *Materials Substitution and Recycling*. Paris: North Atlantic Treaty Organization, 2-1 to 2-16.

_______. 1985. Dire Straits. *Defense and Diplomacy* 3 (no. 8): 16–20.

_______. 1988. *The Structure and Dynamics of United States Government Policymaking: The Case of Strategic Minerals*. New York: Praeger.

Anderson, E. W., and G. H. Blake. 1980. *The NATO Alliance and the Warsaw Pact: Raw Materials and Energy Procurement to 2000: Prolegomena*. London: Ministry of Defence.

———. 1981. *The NATO Alliance and Warsaw Pact: Raw Materials and Energy Procurement to 2000. Selected Nonfuel Minerals*. London: Ministry of Defence.

———. 1982. *The NATO Alliance and the Warsaw Pact: Raw Materials and Energy Procurement to 2000. The Transport of Oil and Strategic Minerals: The Cape Route and Key Choke Points*. London: Ministry of Defence.

———. 1984. *The Republic of South Africa as a Supplier of Strategic Minerals: an Assessment*. Braamfontein, South Africa: Occasional Paper, The South African Institute of International Affairs.

Barnett, F. R. 1981. Reclaiming the Initiative from the Soviet Heartland. The Case for a Tri-Oceanic Alliance. In G. Stewart-Smith *Towards a Grand Strategy for Global Freedom*, 9–24. London: Foreign Affairs Publishing Company.

Barnet, R. J., and R. E. Muller. 1975. *Global Reach: the Power of the Multinational Corporations*. London: Jonathan Cape.

Brandt, W., et al. 1980. *Report on the Independent Commission on International Development Issues*. London: Pan Books.

Bureau of Mines. 1977. *Mineral Perspectives Far East and South Asia*. Washington, DC: Department of the Interior, Bureau of Mines.

———. 1979a. *Mineral Perspectives Mineral Industries of the Middle East*. Washington DC: Department of the Interior, Bureau of Mines.

———. 1979b. *Mineral Perspectives Mineral Industries of Canada, Australia and Oceania*. Washington, DC: Department of the Interior, Bureau of Mines.

———. 1980. *Mineral Facts and Problems 1980*. Washington, DC: Department of the Interior, Bureau of Mines.

———. 1981. *Mineral Perspectives Mineral Industries of Latin America*. Washington DC: Department of the Interior, Bureau of Mines.

———. 1983a. *Minerals Yearbook Vol. 3. Area Reports: International 1982*. Washington DC: Department of the Interior, Bureau of Mines.

———. 1983b. *The Domestic Supply of Critical Minerals*. Washington DC: Department of the Interior, Bureau of Mines.

———. 1984. *Mineral Perspectives Mineral Industries of Africa*. Washington DC: Department of the Interior, Bureau of Mines.

———. 1985a. *Mineral Commodity Summaries 1985*. Washington

DC: U.S. Department of the Interior, Bureau of Mines. Annual issues 1979 to 1986 quoted.

———. 1985b. *Project 2000 A Strategic Plan for the Bureau of Mines.* Washington DC: Department of the Interior, Bureau of Mines.

Carroll, B. A. 1968. *Design for Total War: Arms and Economics in the Third Reich.* The Hague: Mouton.

Central Intelligence Agency. 1984. *The World Fact Book Nineteen Hundred and Eighty-Four.* Washington, DC: Central Intelligence Agency.

Chynoweth, A. G. 1976. Electronic Materials: Functional Substitutions. *Science* 191: 725–32.

Cohen, S. 1964. *Geography and Politics in a Divided World.* London: Methuen.

———. 1973. *Geography and Politics in a World Divided.* New York: Oxford University Press.

Congressional Budget Office. 1983. *Strategic and Critical Nonfuel Minerals: Problems and Policy Alternatives.* Washington, DC: Congressional Budget Office.

Congressional Research Service. 1982. *U.S. Economic Dependence on Six Imported Strategic Non-Fuel Minerals.* Washington, DC: Subcommittee on Economic Stabilization, House Committee on Banking, Finance and Urban Affairs.

Couper, A. D., ed. 1983. *The Times Atlas of the Oceans.* London: Times Books.

Crowson, P. 1984. *Minerals Handbook 1984–85.* London: Macmillan.

Department of Commerce. 1981. *Critical Materials Needs of the U.S. Aerospace Industry.* Washington, DC: Department of Commerce.

Department of Defense. 1984. *M-1 Tank.* Internal paper.

Department of State, Office of the Geographer. 1969. *Sovereignty of the Sea.* Washington, DC: U.S. Department of State, 22–28.

De Vore, C. 1981. Strategic Minerals: A Present Danger. *Signal* 35 (January): 63–68.

Eckes, A. E. 1979. *The United States and the Global Struggle for Minerals.* Austin: University of Texas Press.

Federal Emergency Management Agency. 1986. *Stockpile Report to the Congress* (October 1985–March 1986). Washington, DC: Federal Emergency Management Agency.

Fourie, B. G. 1983. *The Mineral Resources War.* Address to the District of Columbia Bar. Washington DC: 27 January.

GJW Government Relations with Stephenson, P. 1981. *Handbook of World Development.* Harlow: Longman.

Glassner, M. I. 1978. The Law of the Sea. *Focus* 28 (no. 4): 1–24.

Gray, C. 1977. *The Geopolitics of the Nuclear Era: Heartland, Rimlands and the Technological Revolution.* New York: Crane, Russack.

Grichar, J. S., R. Levine, and L. Nahai, 1981. *The Nonfuel Mineral Outlook for the U.S.S.R. through 1990.* Washington, DC: Department of the Interior, Bureau of Mines.

Hanks, R. J. 1981. *The Cape Route: Imperiled Western Lifeline.* Cambridge, Mass: Institute for Foreign Policy Analysis.

Hargreaves, D., and S. Fromson. 1983. *World Index of Strategic Minerals.* Aldershot: Gower.

Hayward, Admiral T. 1979. Address to Sea Power subcommittee of House Armed Services Committee. 20 December.

Holt, J. L., and T. W. Stanley. 1983. *Nonfuel Mineral Stockpiling Policies: the Private Stockpiling Alternative.* Washington, DC: International Economic Studies Institute.

Hydrographer of the Navy. 1980. *Africa Pilot* 3: 294.

Kjellen, R. 1917. *Der Staat als Lebensform.* Berlin: K. Vowinckel.

Landsberg, H. H. 1982. *Minerals in the Eighties: Issues and Policies.* Oak Ridge, TN: Oak Ridge National Laboratory.

Legvold, R. 1983. The Strategic Implications of the Soviet Union's Nonfuel Mineral Resources Policy. *The Journal of Resource Management and Technology* 12 (no. 1): 47–56.

Lloyd's Register of Shipping. 1984. *Lloyd's Annual Report.* London: Lloyd's of London Press.

———. 1985. *Lloyd's Statistical Tables 1984.* London: Lloyd's of London Press, 1–79.

Mackinder, H. J. 1904. The Geographical Pivot of History. *Geographical Journal* 23: 421–37.

———. 1919. *Democratic Ideals and Reality.* London: Constable.

———. 1943. The Round World and the Winning of the Peace. *Foreign Affairs* 21 (no. 4): 595–605.

McMichael, R. D. 1981. Strategic Minerals: Life Subsystem of Resource Dependency. In *Seapower: The Nation's Lifeline Today and Tomorrow.* Pittsburgh: World Affairs Council, 25–31.

Mahan, A. T. 1890. *The Influence of Sea Power upon History 1660–1783.* Boston: Little, Brown.

Mangone, G. J., 1984. *American Strategic Minerals.* New York: Crane, Russak.

Maull, H. W. 1984. *Raw Materials, Energy and Western Security.* London: Macmillan.

May, E. R. 1973. *Lessons of the Past: The Use and Misuse of History in American Foreign Policy.* New York: Oxford University Press.

Meadows, D. H., D. L. Meadows, J. Randers, and W. W. Behrens. 1972. *The Limits to Growth*. New York: Potomac Associates.

Merritt, A. H. L., and L. F. Carter (eds.) 1985. *Mobilization and the National Defense*. Washington, DC: National Defense University, 1985.

Morgan, J. D. 1985. *Minerals: Essential to the Economy, National Security and Standard of Living*. Washington, DC: Department of the Interior, Bureau of Mines.

Muir, R. 1975. *Modern Political Geography*. London: Macmillan.

National Commission on Materials Policy. 1973. *Material Needs and the Environment Today and Tomorrow*. Washington, DC: GPO

National Commission on Supplies and Shortages. 1976. *Government and the Nation's Resources*. Washington, DC: Government Printing Office.

Office of Technology Assessment. 1979. *Management of Fuel and Nonfuel Minerals in Federal Lands. Current Status and Issues*. Washington, DC: Office of Technology Assessment.

———. 1985. *Strategic Materials: Technologies to Reduce U.S. Import Vulnerability*. Washington, DC: Office of Technology Assessment.

Office of the Under Secretary of Defense for Research and Engineering. 1981. *Report of the Defense Science Board 1980 Summer Study Panel on Industrial Responsiveness*. Washington, DC: Office of the Under Secretary of Defense for Research and Engineering.

Paxton, J., ed. 1984. *Statesman's Yearbook 1984–85*. London: Macmillan.

Pirages, S. D. 1978. *Global Ecopolitics*. North Scituate, Mass.: Duxbury.

Plummer, A. 1937. *Raw Materials or War Materials?* London: Victor Gollancz.

Policy Coordinating Committee. 1979. *Report on the Issues Identified in the Nonfuel Minerals Policy Review*. Washington, DC: Department of the Interior.

President's Materials Policy Commission. 1952. *Resources for Freedom: A Report to the President*. Washington, DC: GPO.

Public Land Law Review Commission. 1970. *One-Third of the Nation's Land*. A report to the President and to the Congress. Washington, DC: GPO.

Radice, H., ed. 1975. *International Firms and Modern Imperialism*. Harmondsworth, Eng.: Penguin Books.

The Raw Materials Group. 1981. *Raw Materials for the MX Missile*. Raw Materials Report 1 (no. 1): 6–7.

Seabury, P. 1983. Industrial Policy and National Defense. *Journal of Contemporary Studies* 6 II (Spring): 5–15.

Severin, W. 1983. Soviet Nonfuel Minerals: Recent Trends and Prospects. *Journal of Resource Management and Technology* 12 (no. 1): 37–46.

Siegel, M. A. 1981. *Strategic Materials*. West Palm Beach, Florida: Pratt and Whitney Aircraft.

Smith, R. W. 1974. An Analysis of the Strategic Attributes of International Straits: A Geographical Perspective. *Maritime Studies and Management* 2: 88–101.

Spykman, N. J. 1944. *The Geography of the Peace*. New York: Harcourt Brace.

Thomas, J. R. 1986. *Natural Resources in Soviet Foreign Policy*. New York: National Strategy Information Center.

Trager, F. N., and P. S. Kronenberg. 1973. *National Security and American Society*. Lawrence: University Press of Kansas.

Trager, F. N., and F. L. Simonie. 1973. An Introduction to the Study of National Security. In *National Security and American Society*, edited by F. N. Trager and P. S. Kronenberg, 35–48. Lawrence: University of Kansas.

Trout, B. T., and J. E. Harf. 1982. *National Security Affairs*. New Brunswick: Transaction Books.

Van Rensburg, W. C. J., 1985. *America's Dependence on Imported Strategic Minerals*. Author's personal communication.

Van Rensburg, W. C. J. 1986 *Strategic Minerals*. Volume I. Englewood Cliffs, N.J.: Prentice-Hall.

Walters, R. E. 1974. *The Nuclear Trap: An Escape Route*. Harmondsworth, Eng.: Penguin.

The White House. 1982. *National Materials and Minerals Program Plan and Report to Congress*. Washington, DC: Government Printing Office, The White House.

Zoppo, C. E., and C. Zorgbibe. 1985. *On Geopolitics: Classical and Nuclear*. Dordrecht: Martinus Nijhoff.

Index

About the Author

EWAN W. ANDERSON lectures in the Department of Geography, University of Durham, England. He is also a visiting lecturer at the Universities of Newcastle-upon-Tyne and York. He holds a B.A. degree from the University of Oxford and higher degrees from the following universities: Manchester (M.A.), Durham (Ph.D.), Newcastle-upon-Tyne (M.Ed. and Ph.D.), and York (D.Phil.). His three doctorates are respectively in geology, education, and politics.

Dr. Anderson's current research interests are in the fields of geopolitics, particularly concerning resources, refugees, international boundaries, water, defense industries, and risk assessment; development studies; humanistic geography; and the Middle East.

During 1957–59, he served as an officer in the Royal Navy, and he is still a commander in the Royal Naval Reserve. He has been visiting professor at several European and Middle Eastern universities and spent most of 1985 working in the United States. Dr. Anderson has produced over 100 publications.